EVERY LIVING THING

RSA·STR

THE RSA SERIES IN TRANSDISCIPLINARY RHETORIC

The RSA Series in Transdisciplinary Rhetoric is a collaboration with the Rhetoric Society of America to publish innovative and rigorously argued scholarship on the tremendous disciplinary breadth of rhetoric. Books in the series take a variety of approaches, including theoretical, historical, interpretive, critical, or ethnographic, and examine rhetorical action in a way that appeals, first, to scholars in communication studies and English or writing, and, second, to at least one other discipline or subject area.

Other titles in this series:

Nathan Stormer, *Sign of Pathology: U.S. Medical Rhetoric on Abortion, 1800s–1960s*

Mark Longaker, *Rhetorical Style and Bourgeois Virtue: Capitalism and Civil Society in the British Enlightenment*

Robin E. Jensen, *Infertility: A Rhetorical History*

Steven Mailloux, *Rhetoric's Pragmatism: Essays in Rhetorical Hermeneutics*

M. Elizabeth Weiser, *Museum Rhetoric: Building Civic Identity in National Spaces*

Chris Mays, Nathaniel A. Rivers, and Kellie Sharp-Hoskins, eds., *Kenneth Burke + The Posthuman*

Amy Koerber, *From Hysteria to Hormones: A Rhetorical History*

Elizabeth C. Britt, *Reimagining Advocacy: Rhetorical Education in the Legal Clinic*

Ian E. J. Hill, *Advocating Weapons, War, and Terrorism: Technological and Rhetorical Paradox*

Kelly Pender, *Being at Genetic Risk: Toward a Rhetoric of Care*

James L. Cherney, *Ableist Rhetoric: How We Know, Value, and See Disability*

Susan Wells, *Robert Burton's Rhetoric: An Anatomy of Early Modern Knowledge*

Ralph Cintron, *Democracy as Fetish*

Maggie M. Werner, *Stripped: Reading the Erotic Body*

Timothy Johnson, *Rhetoric, Inc.: Ford's Filmmaking and the Rise of Corporatism*

James Wynn and G. Mitchell Reyes, eds., *Arguing with Numbers: The Intersections of Rhetoric and Mathematics*

Ashely Rose Mehlenbacher, *On Expertise: Cultivating Character, Goodwill, and Practical Wisdom*

Stuart J. Murray, *The Living from the Dead: Disaffirming Biopolitics*

G. Mitchell Reyes, *The Evolution of Mathematics: A Rhetorical Approach*

Jenell Johnson

EVERY LIVING THING

The Politics of Life in Common

THE PENNSYLVANIA STATE UNIVERSITY PRESS
UNIVERSITY PARK, PENNSYLVANIA

Library of Congress Cataloging-in-Publication Data

Names: Johnson, Jenell M., 1978– author.
Title: Every living thing : the politics of life in common / Jenell Johnson.
Other titles: RSA series in transdisciplinary rhetoric.
Description: University Park, Pennsylvania : The Pennsylvania State University Press, [2023] | Series: The RSA series in transdisciplinary rhetoric | Includes bibliographical references and index.
Summary: "Explores the question of what is life, and how invocations of life itself can join and divide, horrify and amaze, and may have the potential to inspire a future politics in a world beset by crises"—Provided by publisher.
Identifiers: LCCN 2022039005 | ISBN 9780271094564 (hardback) | ISBN 9780271094571 (paperback)
Subjects: LCSH: Life. | Rhetoric.
Classification: LCC BD435 .J65 2023 | DDC 128—dc23/eng/20220914
LC record available at https://lccn.loc.gov/2022039005

Printed in the United States of America
Published by The Pennsylvania State University Press, University Park, PA 16802–1003

The Pennsylvania State University Press is a member of the Association of University Presses.

It is the policy of The Pennsylvania State University Press to use acid-free paper. Publications on uncoated stock satisfy the minimum requirements of American National Standard for Information Sciences—Permanence of Paper for Printed Library Material, ANSI Z39.48–1992.

For Jodi and Jenna

Contents

Contents

Acknowledgments

All books begin and, if we are lucky, end in intellectual community. My community begins with the students, staff, and faculty of the University of Wisconsin–Madison and the Wisconsin residents who fund the state university system. I have found many homes on this campus, most notably in the department of Communication Arts but also in the department of Life Sciences Communication, the Holtz Center for Science and Technology Studies, and the Disability Studies Initiative. Thanks especially to Rob Asen, Anirban Baishya, Elizabeth Bearden, Kelley Conway, KC Councilor, Caroline Druschke, Rob Howard, Vance Kepley, Jason Lopez, Lori Lopez, Linda Lucey, Lynn Malone, Sara McKinnon, Nicole Nelson, Lynn Nyhart, Allison Prasch, and Ellen Samuels. Many thanks to my brilliant advisees Liz Barr, Amanda Friz, Erin Gangstad, Allyson Gross, Liam Randall, Dominique Salas, Grant Suhs, and Kendra Winchester, who have all taught me so many things. In many ways, this book began in my 2007 Limits of the Human undergraduate seminar at Louisiana State University, where my students and I wrestled with many of the questions in the pages that follow, conversations that continued in my Limits of the Human and Environmental Rhetoric courses at UW-Madison.

UW-Madison's Constellations program provided me with research support through the Mellon-Morgridge Professorship I held from 2016 to 2021. This funding provided summer funding as well as research assistance from Liz Barr, Amanda Friz, Erin Gangstad, Allyson Gross, Caroline Hensley, and Liam Randall: thank you all from the bottom of my heart. UW-Madison's Institute for Research on the Humanities awarded me a residential fellowship that allowed me the time to push the project to completion and the opportunity to receive feedback from a brilliant group of colleagues from many disciplines. In addition to presenting my work at the IRH, I have presented portions of this book at the National Communication Association's annual conference in 2015 and 2018; the Rhetoric Society of America's biannual meeting in 2014 and 2018; the Society for Literature, Science, and the Arts' annual conference in 2013 and 2015; and the Rhetoric, Politics, and Culture weekly colloquium. Portions of chapter 3 were published as "The End of the World, the Future of the Earth: Bioplurality and the Politics of Human Extinction," *Journal for the History of Rhetoric* 23, no. 1 (2020): 30–53, which is available at https://www.tandfonline.com.

Beyond Wisconsin, I am lucky to be part of a broader intellectual community both inside and outside the academy. This project benefited tremendously from conversations with Allie Rowland and Stuart Murray at the 2014 RSA Summer Institute workshop on bioethics and biopolitics. I first learned about identification in Debbie Hawhee's seminar room over twenty years ago, and it has shaped the way that I think about rhetoric ever since. Thanks also to Vanessa Beasley, Catherine Belling, Jeff Bennett, Kevin Browne, Suzy Cerrato, Karma Chávez, Josh DiCaglio, Scott Graham, Kelly Happe, Robin Jensen, Krista Kennedy, Marina Levina, Melissa Littlefield, Lisa Melonçon, Jodie Nicotra, Ersula Ore, Blake Scott, Susan Squier, Robyn Thoren, Louise Whitely, and so many others in rhetorical studies and beyond.

Caroline Druschke, Melissa Littlefield, John Lynch, Sara McKinnon, and Mike Xenos talked through many ideas with me at length and read early drafts of chapters: thank you all, and a special thank you to Krista Kennedy for essential moral support and advice as I finished this project. My eternal gratitude to Christa Olson, who has given me so much feedback on this project over the years. Many ideas in this book are the product of lengthy conversations with Christa over coffee and croissants as we wrote together on Friday mornings, and she rescued this book from my self-doubt more than once. All errors and oversights are, of course, my own.

Catharine Conley, Dorion Sagan, and Kyle Whyte graciously agreed to be interviewed during some of the darkest days of the pandemic—thanks to all of you for your important work and the inspiration it has given me and for your precious gifts of time and attention.

Thank you to the activists in Extinction Rebellion Madison for your friendly, fierce commitment to a habitable planet and a just world.

This book would not exist without Penn State University Press, the Rhetoric Society of America, and the Transdisciplinary Rhetoric book series that they sustain. I am grateful to the two anonymous reviewers of the manuscript, who helped me to clarify some key points and improve the organization of the argument. Thank you to Josie DiNovo, Andrew Katz, and Laura Reed-Morrisson for shepherding the project through the process and to Ryan Peterson, Kendra Boileau, and Michael Bernard-Donals for their editorial support, and a huge thank you to Leah Ceccarelli for her support of this project from the very beginning and her generous mentorship over the years.

And last, immeasurable gratitude to my family for their love and support—especially Jim and Joy Johnson; Jodi, Ryan, Jack, and Grant Carreon; and Jenna, Kevin, Holden, and Harper Clay. Daily walks with Mabel, the world's sweetest and most stubborn hound dog, helped me untangle many ideas as we wound our way through the grass, leaves, and snow of Hoyt Park each day. But, as always, my most profound debt is to Mike Xenos—for our unending conversation and for being the person I live with, in all the ways that we live together.

Introduction | This Thing We Call Life

You'd think that biologists, of all people, would have words for life. But in scientific language our terminology is used to define the boundaries of our knowing. What lies beyond our grasp remains unnamed.

—Robin Wall Kimmerer, *Braiding Sweetgrass*

Ten minutes ago, there was a fruit fly quietly hovering near my bowl of strawberries. *Drosophila melanogaster.* With its wormy cousin *C. elegans, D. melanogaster* is a classic laboratory organism, a model system. Since Charles Woodworth first bred them for scientific use at the turn of the twentieth century, drosophilae have provided an immeasurable contribution to our understanding of life. Research on *D. melanogaster* has been awarded six Nobel Prizes. So much of what we know about basic biological processes, and especially what we know about genetics, comes from hundreds of thousands, maybe millions, of these tiny lives. Researchers have given drosophila genes cheeky names: Tinman (related to heart development), Van Gogh (related to hair swirls), and Hamlet (affects the development of cells descended from IIB progenitor cells, so "IIB or not IIB"). The average life span of fruit flies is about eight to eighty days, depending on the environment and the circumstances. This particular drosophila probably had been buzzing around my kitchen for days. I don't know if it was born there. Maybe it hitched a ride to my house on these very strawberries.

But it dared to land on the edge of my bowl, and I rendered judgment without thinking.

I think you know what happened next.

There is a difference between a dead fruit fly and a living one, but the nature of that difference is famously hard to pin down. The legendary scientist J. B. S. Haldane opened the title essay of his 1947 book *What Is Life?* with a sly dodge: "I am not going to answer this question." Even though Haldane went on to describe life as "essentially a pattern of chemical happenings" identifiable across a variety of organisms, his initial refusal to answer his own question is striking. Haldane later elaborated on his difficulty: "I doubt if it will ever be possible to give a full answer, because we know what it feels like to be alive, just as we know what redness, or pain, or effort are. So we cannot describe them in terms of anything else."[1] For Haldane, the meaning of "life" is found in some indescribable, indivisible aspect of being alive. Life itself slips from his conceptual grasp as if it were a prime number, something that comes as close to the noumenal realm as a scientist can get without blushing.

The life at the heart of this book is the life that Haldane refused to define: life itself, Life-with-a-capital-L. This is not the life of the pro-life movement, whose adherents, despite professing a "culture of life," tend not to protest the development of a wetland area or the clear-cutting of a forest.[2] For the pro-life movement and for many others who center life in their politics, there is an unstated qualifier: "life" is usually shorthand for some version of "human life."[3] The life of this book, in contrast (or, more accurately, in addition), is the thing you have in common with all other humans but also with owls, birch trees, and bacteria quietly living in the furthest depths of the ocean. And yes, it's something I also share with fruit flies, along with a love of strawberries. This life is the life threatened by full-scale nuclear war, mass extinction, and runaway climate change; the ethical conundrum posed by synthetic biology; the life represented by bacteria that may be hiding in the craggy, dusty regolith of Mars; the life we have inherited from the last universal common ancestor, the progenitor of all known forms of life on Earth. As Haldane makes clear, this broader sense of life—life itself—is difficult to grasp, for life itself, writes Michel Foucault, "does not exist, per se; it is an abstraction."[4]

Life is everywhere on Earth—found in the deepest recesses of its crust and in the far reaches of the atmosphere—yet it may not exist independent of the matter it classifies as animate, as organism, as *a* being, in the Western "ontological economy" at least.[5] And yet, writes Richard Doyle, we have an impulse to "[bow] before the ineffability of the vital," unsure of the features on the face of this god but still convinced that piety is demanded of us.[6] *Life does not exist, per se*, yet it is of such incalculable value that philosophers struggle to provide cogent arguments for why life is good—it is just a matter of intuition, its basic premise a given. *Life may not exist, per se*, but nothing is more important. *Life may not exist, per se*, but it must be protected at all costs.

"Action on behalf of life transforms," writes Robin Wall Kimmerer (Potawatomi).[7] This book is an account of what compels that action, the rhetorical forms it takes, and the kinds of transformation it brings into being. Using a variety of cases in which life itself becomes subject to moral consideration and the subject of political action—what I call *vital advocacy*—this book traces what happens when life itself is evoked through arguments on its behalf. I have no desire to look upon the true face of this god, and I join Haldane in his ontological dodge. What is this thing we call life? *I am not going to answer this question.* But if there is an answer in what follows, it is found in the echoes of its asking, a little like watching the trail of a comet you are not sure was there in the first place.

So let's begin by looking up.

The Overview Effect

The border between Earth and space is one hundred kilometers above your head, a terrestrial boundary known as the Kármán line.[8] At this altitude, the atmosphere is not dense enough to support aeronautical flight. The Kármán line thus marks the difference between airplanes and spacecraft and names the distinction between pilots and astronauts.

In 1961, the Soviet cosmonaut Yuri Gagarin became the first human to cross the Kármán line. Three minutes after launch, officers from USSR ground control asked Gagarin for an update. "I can see Earth," he replied in the staticky staccato of early space communication. "I am looking at the clouds. Beautiful, so beautiful!"[9] Gagarin's flight was not the first time that Earth had been viewed from above: rocket-mounted cameras had been returning grainy black-and-white images since 1946. Nor was Gagarin the first Earthling to cross the Kármán line. Preceding him were fruit flies in 1947, a rhesus monkey named Albert I in 1948, a dog named Laika in 1957, and nameless bacteria that were almost certainly the first Earthlings to reach outer space. Gagarin's flight was significant because he was the first human to leave Earth's atmosphere but also because it was the first time the planet had been seen, unmediated, with human eyes and the first time it had been seen in *color*. At the press conference following his return, Gagarin described his first impression of Earth using vivid language: "The color of the sky is completely black. The stars on this black background seem to be somewhat brighter and clearer. The Earth is surrounded by a characteristic blue halo. This halo is particularly visible at the horizon. From a light-blue coloring, the sky blends into a beautiful deep blue, then dark blue, violet, and finally complete black. . . . Circling the Earth in my orbital spaceship,

I marveled at the beauty of our planet," Gagarin remarked. "People of the world, let us safeguard and enhance this beauty, and not destroy it!"[10]

In the years since, many spacefarers have described a similar suite of feelings in response to crossing the Kármán line, a phenomenon that Frank White has termed the "Overview Effect."[11] The experience of "seeing firsthand the reality that the Earth is in space," White explains, "often transforms astronauts' perspective of the planet and humanity's place in the universe. Some common aspects of it are a feeling of awe for the planet, a profound understanding of the interconnection of all life, and a renewed sense of responsibility for taking care of the environment."[12] Seeing the Earth from space is more than just an aesthetic experience, in other words; it also seems to provoke a profound ethical and political response.

This phenomenon is not limited to astronauts. *Earthrise*, a photograph taken by the Apollo 8 astronauts in 1968, and *22727*, the "blue marble" image captured by astronauts on the Apollo 17 mission in 1972, were widely hailed for their power to generate powerful sentiments in viewers, and they are often credited with catalyzing the rapid growth of the environmental movement in the early 1970s.[13] For many people, these photographs offered "seemingly incontrovertible proof that whatever else might separate us, we are all part of one species, forced to live together on the same fragile planet and sharing the same limited resources."[14] Whole-Earth images thus have been heralded as a means for humans to see themselves joined not by "arbitrary signifiers" such as religion, nation, or even species but by an "unalterable presence," the "undeniable *thusness*" of Earth, a rhetorical move that Tobias Boes describes as "planetary mediation."[15] For many astronauts, however, it is clear that the mediator is not just the planet but the planet as a host for life, to the point that it is sometimes spoken of as a living thing itself. "It is all connected," explains the US astronaut Sandra Magnus. "It is all interdependent. You look out the window, and in my case, I saw the thinness of the atmosphere, and it really hit home, and I thought, 'Wow, this is a fragile ball of life that we're living on.'"[16] The Chinese astronaut Yang Liu describes a powerful feeling: "that the earth is like a vibrant living thing."[17] The US astronaut James Irwin expresses a similar sentiment, calling the planet a "beautiful, warm living object [that is] so fragile, so delicate, that if you touched it with a finger it would crumble and fall apart."[18] One does not look down at the planet and see nations, but neither does one see humanity, as Dorion Sagan explains in the interview that follows this chapter. One looks down, and in the undulating swirls of blue, white, brown, and green, it is life itself that shimmers into view.

One of the most remarkable examples of the Overview Effect is found in a speech given by the US astronaut Russell "Rusty" Schweickart to the Lindisfarne Association in 1974. Inspired by Alfred North Whitehead and Pierre Teilhard de Chardin, the Lindisfarne Association was an eclectic group that brought together artists, academics, scientists, and religious figures "devoted to the study and realization of a new planetary culture."[19] From 1974 to 1977, the association hosted an annual conference, which included themes such as "Mind in Nature" (1977), "A Light Governance for America" (1976), and "Conscious Evolution and the Evolution of Consciousness" (1975).

In 1974, the inaugural conference's theme was "Planetary Culture and the New Image of Humanity." Schweickart, who was part of the Apollo 9 mission to test the lunar module, was asked to speak about his experiences in space and what they meant for the future of humanity. But while he was interested in speaking to this "far-out group," as he described them, Schweickart found himself with a terrible case of writer's block.[20] "I never could prepare for the damn talk," he remembers. "I just couldn't ever get anything done on it, couldn't write even a note; I just mentally blocked." When the time for his speech came, Schweickart stepped in front of the audience without any notes, planning to give some canned remarks about the thrill of space flight. He was an astronaut, after all. *Anything* he had to say would be interesting. As he began speaking, however, something strange began to happen. Schweickart felt himself lose control over his words in what sounds almost like a dissociative experience: "I basically listened to myself give that talk," he recalls. "It really all came out, became conscious to me in that talk. I was almost in the audience."[21]

"Well, what should we do this morning?" Schweickart begins the speech, to scattered laughter from the audience.[22] After hemming and hawing, he explains that he'd like to give the audience something close to his "experience" in space, because the experience of one individual has "very little meaning." At this point, Schweickart then shifts his pronouns from first to second person and his tense from past to present, transforming *my experience then* into *your experience now*.

The first half of the speech walks us through the technical aspects of the mission. The astronaut describes the meticulous process of training and preparation and the launch ("somehow it's anticlimactic" from within the actual vehicle, he notes, because "everything looks very much like the simulations"). He mentions first viewing the Earth from space but does not actually describe this moment in detail; in fact, he describes the experience as something close to a cliché. Once the craft stabilizes, Schweickart explains, "you look out the window and you make some comment. Everybody has to make some comment when they see the Earth for the first time. You make your comment, and it's logged. Duly noted. And then it's to work, because you don't have time to lollygag and sightsee. . . . On with the job." The days whiz by, rote and mechanistic: you wake up, you eat breakfast, you put on your space suit, you test the equipment, you solve problems, you go to sleep. Repeat.

Every so often, Schweickart pauses our journey to comment on the significance of what he is doing, as if remembering that his "far out" audience expects more from him than just a story. For example, he describes the process of walking in space as severing "your umbilical to that mother," the spaceship, with echoes of Kubrick's *2001*, released just a few years earlier. But throughout the first half of the speech, he appears to emphasize his *lack* of self-reflection during the mission. The first time he takes us outside the spacecraft, for example, he calls attention to the "sunrise over the Pacific. But don't look at it. . . . You've got forty-five minutes out there."

There was one point in the trip, however, when the frantic pace slowed down, "a stroke of luck" when Dave Scott's camera jammed while the two men were outside taking pictures (fig. 1). Schweickart recalls a precious minute of calm while he waited: "just a moment to think about what it is we are doing." But then the moment passes—in the mission and in the speech—and he drops the audience back into routine. Wake up. Eat breakfast. Put on your space suit. Do your work. Go to sleep. Repeat, until the moment of relief when "splash, you're on the surface of the Atlantic. You're back in humanity again, and it's an incredible feeling." Having returned his audience safely to the surface of the Earth, Schweickart takes a deep breath, and he breathes out a sigh. And as if interviewing himself, he then asks, almost in a whisper, "And what's it all meant?"

And it is here that the speech changes in both content and delivery. The rapid pace and relentless repetition of the first half downshifts in the second to a slow, steady pulse, and each phrase becomes a stanza. Schweickart returns the audience to the moment of the camera jam; he holds us there, floating in space, outside of time, and he allows the significance of that moment to expand. Gazing down on the Middle East, he explains, "you realize that in one glance that what you're seeing is what was the whole history of man for years—the cradle of civilization. And you think of all that history that you can imagine, looking at that scene." Then it's North Africa that comes into view, then the Indian Ocean, then the Philippines, then the "monstrous Pacific Ocean," and then "you finally

Fig. 1 | Photograph taken by Russell Schweickart from the porch of the Apollo 9 Lunar Module, March 3, 1969. NASA.

come up across the coast of California, and you look for those friendly things: Los Angeles and Phoenix and on across El Paso, and there's Houston, there's home, you know, and you look, and sure enough there's the Astrodome. You know? And you identify with that, you know—it's an attachment." The closing of Schweickart's speech extends this point, and it rewards a slow reading:

> *That identity!* You identify with Houston, and then you identify with Los Angeles and Phoenix and New Orleans and everything. And the next thing you recognize in yourself is you're identifying with North Africa. You look

> forward to that; you anticipate it. And . . . there it is. And *that whole process begins to shift what it is you identify with.* When you go around it in an hour and a half, you begin to recognize that your identity is *with that whole thing.* And that makes a change. . . . You look down and see the surface of that globe you've lived on all this time, and you know all those people down there, and they are like you, they *are* you—and somehow you represent them. You are up there as the sensing element, that point out on the end. That's a *humbling* feeling. It's a feeling that says you have a responsibility, it's not for yourself. . . . Somehow you recognize that you're a piece of this total life, and you're out on that forefront, and you have to bring that back, somehow. And that becomes a rather special responsibility, and it tells you something about your relationship with *this thing we call life.* . . .
>
> And when you come back, there's a difference in that world now; there's a difference in that relationship between you and that planet and you and all those other forms of life on that planet, because you've had that kind of experience. And it's a difference, and it's so precious. And all through this, I've used the word "you" because it's not me—it's not Dave Scott, it's not Dick Gordon, Pete Conrad, John Glenn—it's you, it's us, it's we, it's *life that's had that experience.* . . . And I guess that's really about all I'd like to say, except that—and I don't even know why, but to me it means a lot—and I'd like to sort of close this part of it . . . with, um, uh, a poem, a poem by e. e. cummings that has just become a part of me, somehow out of all this, and I'm not really sure how.

I have listened to this speech dozens of times now, and when Rusty Schweickart reaches its conclusion, a poem by e.e. cummings thanking God for the "greenly spirits of trees," I often find that I've been holding my breath. Maybe it's his stream-of-consciousness delivery, described by one audience member as a "long, pauseless prayer."[23] Maybe it's the striking use of the second person and present tense to create the Overview Effect in his audience, an immersive, evocative *phantasia.* Maybe it is listening to this fighter pilot, engineer, and astronaut wax transcendental at the recognition that he is the "sensing element" of "this thing we call life." Seeking to communicate the meaning of his time above the Kármán line, removed from the terrestrial geography of nearly all human history, Schweickart's off-the-cuff speech transforms to prose and dissolves into poetry in the attempt to communicate his sublime experience. *Sub limen.* Beyond the line.

Bioidentification

Rusty Schweickart's Lindisfarne speech is an exemplar of the rhetorical phenomenon at the heart of this book, which I call *bioidentification*. Bioidentification names the evocation of life as a shared substance, as well as the feeling that such connection produces.[24] I build this idea from Kenneth Burke's theory of identification, one of the most well-known concepts in rhetorical studies. Presented first in *Rhetoric of Motives*, the second volume in Burke's intended trilogy of rhetorical theory, identification is defined as a rhetorical act in which audience members are moved by something they have in common or are led to believe they have in common: a rural upbringing, a passion for justice, an alma mater, a national identity. You persuade someone, Burke writes, "insofar as you talk his language by speech, gesture, tonality, order, image, attitude, idea, *identifying* your ways with his."[25]

While the classic example of identification is the evocation of a commonality between a rhetor and audience, it can also be broader and more diffuse. Humans may identify, or be identified, with nonhuman animals or ideas or things.[26] Identification may be created not just between the rhetor and audience but also between the subject of a speech and the rhetor or between the subject and the audience. In an epideictic speech, for example, the audience may be identified with bravery, or the rhetor with kindness.[27] Identification may also be found in the circulation of discourse, in which ideas, affects, terms, values, and symbols—a "body of identifications"—stick to each other, to people, and to institutions through proximity and accumulation.[28] Writing shortly after the nuclear devastation of Hiroshima and Nagasaki, for example, Burke points to the militarization of science and technology to show how identification may operate beyond intention. When science is identified with the military, scientists are also identified with the "moral qualities" or the "motives" of the latter, Burke argues, no matter their protest to the contrary.[29]

What I am calling *bioidentification* is a simple idea. It's not a transformation of Burke's concept but simply one form that it takes. To see how it works, let's return to Schweickart's Lindisfarne speech. Gazing down on Earth, Schweickart describes how his identity (and, through the explicit use of second person, his audience's identity) does not just shift but expands—from Houston, the place where his friends and family are, to other US cities, then to other countries, then to other continents, until he identifies with Earth itself, with "that whole thing." The people below, he explains, are "like you," but then he corrects: "they *are* you"—yet he does not mean that we are *identical* to them. Nor do we merely resemble each other. Rather, it's that we share something fundamental in common. Life, for which he serves not only as a representative but as the "sensing element," is the thing that identifies Schweickart with all people and with the "greenly spirits of trees" but also with "all those other forms of life on Earth."

Schweickart comments that his experience of bioidentification is "a humbling feeling," which brings to mind Kant's description of humility as a "*sublime* temper of the mind."[30] Thinkers have long described the sublime as a feeling of pleasure, wonder, and terror provoked by an encounter with greatness, which can take the form of power, size, or overwhelming beauty. We encounter the perfect work of art, the voice of the divine, or the limitless universe, and we shake in their shadow. But this typical description of the sublime is incomplete, write Joshua Gunn and David Beard, without noting the "unsettling" shift in subjectivity that accompanies such an experience, in which "the subject is revealed to be a fragile, incomplete construction rather than an integral whole."[31]

There's no question that Schweickart experiences a shift in subjectivity during his time in space, but that doesn't quite account for what is going on in this speech. In *Scale Theory*, Joshua DiCaglio suggests that the rhetorical power of the Overview Effect might be found in its capacity to elicit a "scalar experience" in the viewer. He illustrates this concept using the example of his own encounter with an image of the whole Earth:

> [I] know that I perceive it, but—I seek for another means to say it but can only go to a scalar comparison to describe the experience—it is like saying I see the atoms in my hand when I look at my hand. My mind reels at the suggestion, experiencing both scales at once for a moment. *All of this here, is in that orb.* And this orb is seamless, vibrant and clear in its spherical vastness. Now the duck-rabbit switch is my existence itself, felt whole, divided, whole, divided. . . . I feel suddenly as if I speak for the Earth, like my hands type for it. . . . I can no longer tell if I am seeing the image or speaking to it—I sidle around its surface—I feel its emanation and feel how I emanate from it. I feel lost in its surface but still embedded, strikingly aware of my feet on the ground, of the pull of gravity—the testament of the immensity of this presence.[32]

Edmund Burke famously argued that the terror of the sublime is linked to our mortality, the ultimate dissolution of the subject.[33] But Schweickart and DiCaglio point to a second movement of the sublime: a scalar shift from the "diminution to the aggrandizement of the subject," as Christopher Hitt explains it.[34] Schweickart's experience of the sublime may be tied to a sense of mortality—his life protected by mere inches of fiberglass cloth—but it also seems to speak to a twinned sense of *vitality*, a profound, sublime connection with "this thing called life."

This phrase returns us to the ontological dilemma with which we began. It also raises the question of what it means to approach life itself as separate from that which it qualifies as *living*; that is, it raises the issue of what it means to consider life as a *substance*. Substance is key to Kenneth Burke's theory of identification even as he admits that the term is "beset by a long history of quandaries and puzzlements."[35] Another word Burke uses for identification, in fact, is "consubstantiation," which speaks to identification as a form of connection, but one that stops short of total union. In one of the most well-known passages in *Rhetoric of Motives*, Burke describes this phenomenon as follows: "A is not identical with his colleague, B. But insofar as their interests are joined, A is identified with B. Or he may identify himself with B even when their interests are not joined, if he assumes they are, or is persuaded to believe so. Here are the ambiguities of substance. In being identified with B, A is 'substantially one' with a person other than himself. Yet at the same time he remains unique, an individual locus of motives. Thus he is both joined and separate, *at once a distinct substance and consubstantial with another*."[36]

For Burke, the etymology of "substance" reveals an instructive paradox at its center. "Sub-stance" literally means *that which one stands upon*. In a similar way, Burke writes, "the word 'substance,' used to designate what a thing *is*, derives from a word designating something that a thing *is not*. That is, though used to capture something *within* the thing, *intrinsic* to it, the word refers to something *outside* the thing, *extrinsic* to it."[37] We often assume that substance speaks to a kind of timeless essence—something that is intrinsic to some *thing*. However, the paradox of substance is that we "only ever encounter the *qualities* of the object, and never the *substance* of the object," explains Levi Bryant.[38] "Each time you scrutinize a concept of substance, it dissolves into thin air," Burke writes. "But conversely, the moment you relax your gaze a bit, it reforms again."[39] *What substance is* is less important to Burke than *what substance does*; that is, substance is significant for the function it serves, the "covert influence" it exerts.[40] He "is less interested in 'substance' than in 'substantiating,'" Weldon Durham explains.[41] Burke approaches substance, in other words, not like a scientist or even a philosopher but like a rhetorician—he is more interested in the symbolic work that substance does in the world, rather than what it is, *really*.

Stare at "life" too long, and it becomes an afterimage. And maybe an afterimage is all it ever was, all it ever is, an imprint with no visual stimulus, a scholarly palinopsia, a "rhetorical black hole."[42] To be frank, the moments when I have tried to look too closely at life itself are also the moments I have nearly abandoned this book. Sometimes it feels like I'm writing about the most important thing in the world, and sometimes it feels like I am writing about nothing at all. Life is a little like consciousness that way, in that it seems to inspire lifelong projects of futile devotion. I sometimes worry that life itself is a kind of holy grail, the pursuit of which famously drives its hunters mad. Then again, maybe a little bit of madness helps keep the scholarly eye unfocused enough to appreciate how life moves, and moves us, without getting distracted by the details.

Vital Rhetoric

In *Experimental Life*, the literary critic Robert Mitchell argues that we are in the midst of a "vital turn," in which scholars have become enamored with "life itself as a source of mystery and provocation."[43] This is not a new phenomenon: the first wave of what Mitchell calls "experimental vitalism" flourished at the turn of the nineteenth century, in which an enchantment with life occupied European scientists, physicians, and authors of Romantic literature. The second wave Mitchell locates at the turn of the twentieth century, when biologists like Hans Dietrich developed theories of life based on embryology and philosophers like Henri Bergson theorized the creative potential of an *élan vital*, thought experiments that would shape philosophy, literature, and art for a generation. In the current moment, the explosion of research in fields like molecular biology, biophysics, biogeology, and astrobiology and the existential tinkering of synthetic biology, gene editing, and artificial intelligence represents a broad scholarly interest in life itself that crosses disciplines and brings new disciplines into being. This turn is visible in several currents of inquiry in the humanities and social sciences, too: actor network theory, new materialism, the ecological turn, the post-human turn, the affective turn, as well as research on process, emergence, systems, and complexity and the library of scholarly literature inspired by theories of biopolitics.

While an exhaustive account of rhetoric's own vital turn is beyond the scope of this book, it might be identified in rhetorical research on some of the topics just listed, as well as in the emergence of critical terms like "bio-rhetoric" (which refers to the circulation of biological discourse within social and political discourses), "biocriticism" (which examines rhetoric at the "nexus of disease and culture"), "biocitizenship" (which examines the intersection of bodies, belonging, and political action), and, of course, biopolitics.[44] However, while life and death have been central themes in rhetorical studies for many years, there has been less attention to life itself as a rhetorical object, or what we might think of as "vital rhetoric."[45] I introduce this term not for the sake of adding yet another neologism to an already bloated bio-vocabulary but to distinguish the life that is the subject of this book from other important ways we might understand it.[46] Vital rhetoric names how life itself is evoked as a substance, separable, even if just rhetorically, from living things—something that comes close, perhaps uncomfortably so, to vital*ism*.

While vitalism takes several forms across history, Byron Hawk points out that all vitalisms share something fundamental in common, something I refer to in these pages as the grail question: *What is life?*[47] Ask a physicist what life is, and they might tell you that it is a temporary resistance to entropy. Ask a biologist, and they might talk to you about metabolism or autopoiesis. Ask a priest, and you might get talk of the soul or the Holy Ghost. Ask this rhetorician, and she will answer that whatever else it may or may not be, *life itself is a kind of rhetoric*. Make no mistake: I am not saying that life itself *is* rhetoric. When a fruit fly is crushed, I am not sure exactly what ceases to be, but I know that it's something more than rhetoric. But, like Burke's approach to substance, a rhetorical perspective invites us to turn away from questions of what life is and toward questions of what "life" does: what it joins together and what it divides, what it makes possible, what it prevents, and what consideration, value, and protection it guarantees.

I felt a little guilty for killing that fruit fly. Did you judge me, just a little bit? We are taught to revere life as not just valuable but invaluable, *something special*, and this specialness commands a right response. E. O. Wilson, that great aficionado of insects, has described a version of this response as "biophilia," the "innate tendency to focus on life and lifelike processes."[48] For Wilson, biophilia has an ethical dimension: life has a moral weight that seems to demand a new way of thinking (for a Western scientist, at least) about our relations and obligations to the living world. "It is time," he writes, "to invent moral reasoning of a new and more powerful kind, to look to the very roots of motivation and understand why, in what circumstances and on which occasions, we cherish and protect life."[49]

Note the contingency in Wilson's call for a moral consideration based on life itself, or what is sometimes called "biocentric ethics." As we will see in chapter 1, the meaning of "life" depends on the context in which it is evoked and understood. It is thus linguistically, culturally, and historically specific—and so are the moral frameworks in which it is ascribed value. In the acclaimed book *Braiding Sweetgrass*, for example, the environmental scientist Robin Wall Kimmerer describes how learning the language of her Potawatomi ancestors shifted her understanding of the life at the center of her work. While Kimmerer's training had sharpened her skill at observation, she writes, its language, a "language of distance," a "language of objects," was "based on a profound error in grammar": "My first taste of the missing language was the word *puhpowee* on my tongue. I stumbled upon it in a book by the Anishinaabe ethnobotanist Keewaydinoquay, in a treatise on the traditional uses of fungi by our people. *Puhpowee*, she explained, translates as 'the force which causes mushrooms to push up from the earth overnight.' As a biologist, I was stunned that such a word existed. In all its technical vocabulary, Western science has no such term, no words to hold this mystery."[50]

Kimmerer's describes the Potawatomi language as a lively grammatical universe in which rocks and grandmothers are both "whos," a linguistic realm with shades of nuance that neither science nor English has the capacity to express. Kimmerer explains that the Potawatomi grammar of animacy, which is linked to an understanding of kinship between human and other-than-human worlds, has profound ethical and political implications. It is possible that thinking of a tree as a "she" makes it harder to chop her down, she writes. "Maybe a grammar of animacy," she suggests, "could lead us to whole new ways of living in the world, other species a sovereign people, a world with a democracy of species, not a tyranny of one; with moral responsibility to water and wolves."[51]

It is tempting to suggest that the moral consideration Kimmerer grants to water and wolves exemplifies the relationship between bioidentification and vital advocacy that this book traces. However, such a claim is troublesome, a point I return to in chapter 1. I wish neither to claim that the Potawatomi grammar of animacy is *really* a form of bioidentification nor to extract this system from its original context, stripping its subtlety to suit my purposes. Kimmerer's experience learning *puhpowee* teaches me that any attempt to examine the questions of life, ethics, rhetoric, and politics around which this book orbits reflects the language, time, and culture in which those questions are asked. It also reminds me to consider the relationships of power and knowledge that give meaning and authority to the answers. *Puhpowee,* in other words, is not the missing answer to J. B. S. Haldane's question, and it is not the thing that Rusty Schweickart claims to be representing while gazing on the Earth from space. *Puhpowee* is not the true name for "life" any more than "life" is the true name for the force that pushes mushrooms up from the ground. *Puhpowee* is the word that the Potawatomi people found for *puhpowee,* and its place in this conversation ought to be on its own terms.

Chapter Overview

The thread that draws the chapters of this book together is *vital advocacy:* rhetorical action on behalf of life itself. The advocates in the following pages are a rather motley crew: scientists and astronauts and philosophers and activists and provocateurs and science fiction writers. Some you may see as heroes, and some you will almost certainly see as villains. I chose the case studies for their resonance but also their dissonance, which offers instructive points of tension that give this book's conversation energy and, I hope, wider significance. While the rhetorical perspective of the book draws my attention to symbols, tropes, language, meaning, and argument in its various forms, bioidentification is the term at the center of the analysis, and I return to the ethical questions and political imperatives it raises about similarity and difference, self and other, and connection and division, again and again.

As I noted earlier, the ultimate question this book asks is not *What is life?*—the grail question with which we began—but *What does "life" do?* The chapters that follow are a collection of answers to that question, and they are arranged to produce an evocative, echoing, lively conversation that adds interlocutors as it proceeds, rather than a teleological argument with a pat conclusion. In that dialectical spirit, I have included three interviews with individuals who have inspired my thinking in this book: Dorion Sagan, who has written widely on the topic of life, including a number of books coauthored with his mother, the famed microbiologist Lynn Margulis; Kyle Whyte (Potawatomi), a scholar working at the intersection of environmental and climate justice, ethics, and Indigenous studies; and Catharine Conley, an astrobiologist who served as NASA's Planetary

Protection Officer for over a decade. These conversations stand alone between chapters, and each addresses themes of the preceding chapter while also raising new lines of inquiry. How, Sagan points out, does our grammar for life, as noun or as verb, shape our understanding of it? What, Whyte asks, does kinship mean for our notions of responsibility and consent? What, Conley invites us to consider, would it mean to discover life elsewhere, and what steps should we take to protect it?

The book begins by defining bioidentification by identifying its epistemic location and defining its limits. In chapter 1, I identify the Western "locus of enunciation" of Burkean identification, focusing on a priori division that identification seeks to bridge.[52] To identify Burke as a white Eurocentric writer or to name rhetoric as a discipline with a Western bias is the ripest fruit on the lowest branch; instead, the question this chapter asks is, Why does this matter? What does it mean to think about life itself as a Western idea? What do these locations allow us to see? What is hidden from view? The chapter then offers an "anti–case study" of the Lakota phrase *Mni Wiconi* (water is life), which achieved wide circulation during the efforts to halt the Dakota Access Pipeline (DAPL) in 2016–17. *Mni Wiconi*, and the Lakota cosmology it emerges from, reveals an approach to life not as thing but as a capacity, which binds not only humans and other humans, and humans and nonhumans, but also biotic and abiotic worlds. There is a lesson to be learned in examining the difference between Indigenous and Western perspectives on life, as the philosopher Brian Burkhart (Cherokee) argues, and in thinking through the misfit between them.[53]

Chapter 2 takes a deep dive into deep ecology, an approach to environmental thinking, ethics, and politics that has inspired a number of radical environmental movements over the years. Instead of viewing deep ecology as a politics or philosophy, I proceed from the idea that it is best viewed as a rhetoric.[54] Focusing on the use of the term "identification" in the writing of deep ecology's founder, the Norwegian philosopher Arne Næss, I show how deep ecologists use life to create consubstantiality with the other-than-human world, a rhetorical strategy that carries a sense of commonality, vulnerability, and obligation across species. While deep ecology's bioidentification has been the source of positive inspiration to many environmental thinkers and activists, I conclude this chapter by showing how its biocentric worldview can lead to dark places, focusing on the writing of Pentti Linkola, a Finnish ecologist whose antihuman rhetoric has been cited as inspirational to the troubling growth of ecofascism in recent years.

Chapter 3 takes up where chapter 2 leaves off, and in some ways, these chapters are two halves of the same argument. What does it mean to understand human life as responsible for the mass death of the living world, a time that some have called the Sixth Mass Extinction? Conversely, what does it mean to imagine the death of humanity in toto as the grounds by which other life may survive? This chapter examines the rhetorical role of human extinction in two social movements—the Voluntary Human Extinction Movement and Extinction Rebellion—to consider the promise and pitfalls of gathering humanity under the banner of species. This chapter instead suggests an approach of *bioplurality*, building on insights from Hannah Arendt and Sylvia Wynter, which highlights the twinned role of identity and difference in creating solidarity among and within species.

Using the backdrop of planetary belonging, chapter 4 considers our obligations to life on Earth, understood as life *on Earth*, as well our responsibility to potential life elsewhere in the universe. Using the "humble microbe" as its artifact, and the issue of interplanetary contamination as its exigence, this chapter examines how controversies about interplanetary contamination in space policy, space science, and science fiction turn on arguments of magnitude that is given meaning by place. Beginning with the emergence of the field of exobiology in the 1950s, debates about biological threats to Earth and other celestial bodies, and concluding with the imagined terraforming of Mars, this chapter explores how relations between planets are understood as an existential threat, which can take the form of invading Martians or a mere microbe on a rover's wheel.

The book concludes with a final brief case study, a reading of a recent essay by the political theorist Achille Mbembe, which draws out the stakes of vital advocacy in a moment beset by interlocking planetary crises. Mbembe has become well known as an influential theorist of death—at last count, his essay "Necropolitics" has been cited over eight thousand times. In this brief conclusion, I examine how the bioplurality in Mbembe's essay reveals a radical politics of life beyond the technique of control and management identified by theories of biopolitics and necropolitics. What we see in Mbembe, I argue, is the glimmer of a future vital politics, whose task is nothing less than changing the world.

What draws scholars together in this current vital turn, Mitchell argues, is an approach to life not as "a source of perplexity that demands new modes of conceptual and practical experimentation."[55] In these experiments, life itself takes shape as something like an aporia—a room with no door, a conundrum without a clear answer. "If we arrive at an aporia," writes Stuart Murray, "it means we are in doubt, we are perplexed, we are confused about how (best) to proceed. An aporia is a contradiction, a puzzle or a paradox." An aporia may produce an impasse in thinking, but it also demands that we shift our tactics of inquiry: the questions we ask as well as the places we go to for answers. Aporias, Murray explains, following Derrida, reveal the limits of science and logic, but in so doing, they also "deliver us over to the ethical and the political," where "new ways of speaking and thinking and relating" may be found.[56] There is not an answer to Haldane's question in what follows, or at least not one that would satisfy a scientist like him. Confronted by the aporia *What is life?* we, like Haldane, cannot answer, but neither can we walk away; rather, as Derrida writes, "we are going to wander about in the neighborhood of this question," examining the rhetorical, ethical, and political terrain along the way.[57] What follows, then, is my own experiment born of wandering: a rhetorical account of life itself through a multiplicity of voices telling stories about what it means, what it does, and why it matters.

Life Is Like a Verb | A Conversation with Dorion Sagan

Dorion Sagan is a celebrated writer, ecological philosopher, and author or coauthor of twenty-five books, which have been translated into fifteen languages. As an ecological theorist, he has been at the forefront of bringing our growing understanding of symbiosis as a major force in evolution into the intellectual mainstream, within both science and the humanities, and rethinking the human body as a "multispecies organism." Sagan is a serial collaborator on scientific, intellectual, and artistic projects, and his work ethic follows that of evolving life, whose creativity derives largely from symbiotic merger and genetic recombination. With Carl Sagan and Lynn Margulis, his parents, he is coauthor of the entries for both "Life" and "Extraterrestrial Life" in the *Encyclopedia Britannica*.

JJ: In *What Is Life?*, your book with Lynn Margulis, you offer one of the most elegant definitions of life that I've come across. You describe the question "What is life?" as a "linguistic trap. To answer according to the rules of grammar, we must supply a noun, a thing, but life on Earth is more like a verb."[1] I love that line. Why is it better to think about life like a verb? What does it help us to understand?

DS: The book was a collaboration, and at this point, it's hard to say who wrote which part. But that part has often been quoted, "life is a verb," which I think isn't quite as good as "life is *like* a verb." Because the word "verb" is a noun, too. There's only one example we know of life as a contiguous phenomenon, and that's the example of which we are a part. This is the basis from which we make our observations of that which is not life, which, because of that entanglement, always already implicates whatever it is we're trying to describe. We're actually just trying to describe ourselves from the outside, from the inside, when we say that. So, already you see there's a big mélange of levels in the very process of trying to talk about it.

So this example is not fully circumscribed by describing it as a vitalistic substance but is better viewed as an ongoing process—one that is evolving. This unique known example, life on Earth, appears to have a local history of about four billion years. During those four billion years, it's gone from using just a few elements in the periodic table of elements, including carbon, hydrogen, nitrogen, oxygen, phosphorus, and sulfur, and the proteins and nucleic acids of early cells to being a process that has involved ever more substances. You have silicon being mined from the oceans to make the silica shells of diatoms. You have carbon in the atmosphere of the Earth being taken and put into the ground by photosynthesis on a massive scale. The fossil record shows that life spread from water to land and that it oxidized the atmosphere, because there are uranium oxides and iron oxides on the land dated from about two billion years ago, which are attributed to the cyanobacteria, which used the hydrogen of water in photosynthesis, releasing molecular oxygen as a waste gas. The material details of what we call life continue to change. We're talking about a four-billion-year process, which hasn't stopped, and it is much different now than it was before.

If I, for example, were to define you on the basis of your dimensions, the various colors you're wearing right now, and say, "That's you," well, that would not necessarily match with who you were as a zygote, right? Let alone the preprocess of your life through your ancestors! It would give you a false sense of finality to define you as a life form as a thing. To go back to my earlier comment about the increasing number of chemical elements that have become involved in life over evolutionary time, we wouldn't want to just define it as a material thing that just consists of this element and that element and that one. Because what it appears to be is an expanding, evolving process that integrates metabolism and evolution over thousands of millions of years and is growing and becoming more complex and developing more skills and commandeering more energy and spreading more energy and finding more ways of metabolism and more complex structures, more types of cells, and on and on. It's a process, and it's a certain kind of process. To call it a "thing" would be to short shrift it.

JJ: I love that.

DS: That idea of adding more and more chemical elements comes from Vladimir Vernadsky, who popularized the term "biosphere." The reason I wanted to mention Vernadsky is because in Russian, he apparently avoided the term "life" and preferred to say "living matter" and also used the term "animated water," which life mostly is compositionally.[2] In any case, it's dangerous to say "life" if you want to try and understand it. Indeed, as the poet Rilke pointed out, it's dangerous to give anything a name because you might then associate the thing or the process with the name and then no longer see it with an open mind. Because of the history of science and the idea that life is this animate, enchanted, God-made thing that's separate from the environment or a spirit that's stuck in matter etcetera, to even use the word "life" flirts heavily with vitalism. By wrapping a four-letter word around this process, you are making the implicit assumption that you know what it is and that it is a stable thing. But it's not. It's a kind of slow-motion storm transmuting the energy of the sun through photosynthesis into the room-temperature burning of the oxygen-based metabolism of surface organisms. So, there's a trick even in the question "What is life?" It is not Platonic but Heraclitian, changing.

JJ: In Robinson Jeffers's poem "The Treasure," he describes life as a "found word," a "noise" for a kind of reverent silence.[3]

DS: I love Robinson Jeffers. He was one of Bukowski's favorite poets, too.

JJ: Speaking of sublime things, in the introduction of *What Is Life?*, you quote astronaut Eugene Cernan, who describes an ecstatic vision of the Earth from space. Why do you think that people turn to the Overview Effect—the experience of seeing life from space—when they're trying to capture something about the ineffability of life? Why, when trying to explain this thing we call life, do so many people turn to the Overview Effect?

DS: I've written about this in an essay called "What Narcissus Saw." I argued that what was going on when they came back in the '60s with the first images of the Earth from space wasn't so much that we were seeing the Earth. What was happening was that we were seeing ourselves. It's as if we were Narcissus in the myth. In the myth, he hears Echo in the background, who repeats the end of his sentences and is also a personification of his own voice. Then, above the water, he falls in love with his image, which he's never seen before, and he drowns. So, this being spellbound by oneself can be a dangerous process. But I think that it's a good metaphor for the power of that image. It's not just that we see what looks like a spherical being, as well as a planet. When you see the swirling clouds in those pictures, their shape and substance is influenced by water using and transporting life. The processes of life go to the deep biosphere and into the atmosphere, which is continuously produced by life, for example, by those little diatoms, which may account for up to 40 percent of the annual production of oxygen we breathe.

There are major, mostly abiogenic processes like the Hadley cells in the atmosphere that are dissipating heat. But some processes, including mountain formation, may be mediated by the activity of life on Earth's surface. There were plenty of oceanic algae before plants evolved or animals followed them to land. For example, there are microalgae that, when virally infected or when they become too populous, undergo apoptosis, so-called program cell death, and they sink in submarine rain to the bottom of the ocean. They may become lubricant as calcium carbonate for tectonic plate action, increasing volcanic activity and mountain building. As far as I know, Earth has the most active tectonic-volcanic regions in the solar system. So, maybe the extensive nature of life beyond our bodies and into the body of the Earth is something that we intuitively detect when we see this picture of swirling life from space. It's literally an ecstatic experience of being outside yourself.

JJ: In Rusty Schweickart's speech to the Lindisfarne society, there's this beautiful moment when he describes himself as "the sensing element of life." I wonder if that's kind of what he was getting at with that, and not just life now—life *until* now.

DS: He's definitely a good avatar of the best part of space exploration, I think.

JJ: Okay, so, let's turn from very big to very small things. We're talking to each other in the middle of a pandemic. I know that in *What Is Life?* you and Lynn Margulis take the position that viruses are not alive. I was wondering if you could say a little bit about why that is?

DS: Well, a virus, like a gene, does nothing by itself. It does not have a metabolism. According to my mother, the minimal unit of life is the bacterium or prokaryotic cell. Viruses have a lot of activity but only within a metabolizing being. It's kind of like your car in your driveway. It may get you to a party. You may meet somebody. You may eventually reproduce with that person. Insofar as you've done that, the car was part of the reproductive process. But the car by itself isn't going to do anything. It's like a machine or meme, involved in its own replication like the words "copy me," but it's a fragment, unable to do anything on its own. Whereas even a microbial cell, a bacterium, if it fulfills its energetic and material needs, continuously produces itself in a process that is sometimes called autopoiesis, self-making. Viruses may have originated as fragments of DNA or RNA and protein that got separated from functioning autopoietic cells. It's also possible that life evolved from a milieu that contained a lot of genetic fragments and that viruses or virus-like particles were part of the backstory of the origins of life.

By themselves, viruses are only potentially active, replicating strands of DNA or RNA that are covered in protein. But they have to get into one of these circular, continuously self-producing, living systems. In that sense, they're no more living than a corpse. The difference between a bull when it's attacking a matador and the bull when it's dead on the dirt is a very big difference, although there's not much material difference. The difference is that it's no longer autopoietic; it's lost its metabolic circularity, its informational closure, the self-reflexivity both physical and sensorial that makes a being alive. It was once alive. With a virus, it's more like the car. It was never alive by itself.

JJ: It seems that people want to attribute agency to viruses, or independence. It reminds me of a line in *What Is Life?* where you describe independence as a political term, not a scientific term.[4]

DS: That's sort of the flip side of that argument. Since that book came out, there have been studies saying that the evolution of the placenta would be impossible without viruses and that up to, or possibly over, 50 percent of the human genome has a viral origin. So, viruses are definitely involved in the process of life. It's a very complex negotiation. The silicon-mining diatoms, the calcareous algae I told you about earlier? These beings taking carbon from the atmosphere and giving off oxygen and sometimes getting deathly sick from viruses or overcrowding and falling to the floor of the ocean—such beings with their naturally evolved nanobiotechnologies are *also* involved in the circulation of calcium and sulfur and other elements. They are mediators in ocean temperatures, volcanic eruptions, and formation of continental shelves and ocean floor. The tiny and contagious are involved in the planetary and stable.

So, too, death is integrated into life. It's not the Manichaean either/or-ness of the germ theory of disease where bacteria and viruses are always bad. You have a continuous multibillion-year negotiation of diversifying life-forms, some of which kill each other and some of which save each other by killing each other. Bacteria are killed by bacteriophages, and that's not necessarily a bad thing, right? The *Shigella flexneri* that can cause dysentery are also killed by these "bacteria-eating" viruses. And bacteriophages are the most common form of encapsulated nucleic acid on Earth, by some estimates more prevalent than all full organisms and viruses combined. And they might save your life.

1

Life in Water, Life in Stone | The Limits of Bioidentification

Life is fundamentally the capacity for kinship.
—Brian Burkhart, *Indigenizing Philosophy Through the Land*

In September 2014, representatives from the Standing Rock Sioux Tribal Council met with representatives from Energy Transfer Partners (ETP; now Energy Transfer), the corporation planning a pipeline that would transport shale oil from the Bakken oil fields of North Dakota to a terminal in central Illinois. An earlier proposal that would have sited the Dakota Access Pipeline (DAPL) ten miles north of Bismarck, whose population is 93 percent white, was rejected by the Army Corps of Engineers due to the risk to Bismarck's municipal water supplies.[1] The revised plan called for the pipeline to cross Mni Sose (the Sioux name for the Missouri River) just half a mile north of the Standing Rock reservation. Standing Rock Tribal Chairman David Archambault III opened the testimony by framing the DAPL as a violation of tribal sovereignty and treaty boundaries: "So, just so you know coming in," he told ETP representatives, "this is something that the tribe is not supporting. This is something the tribe does not wish." Other council members described in detail how the pipeline posed a grave risk to their water supply and also threatened several sites of cultural and historical significance to the Oceti Sakowin (the Dakota, Nakota, and Lakota people).[2]

Near the end of the meeting, Standing Rock Councilwoman Phyllis Young took the microphone and addressed the ETP representatives with a resolute voice: "You know, we have survived incredible odds. You must know that we are Sitting Bull's people. You must know who we are. And we sit on a previous military fort, and we have survived that. And we have rebuilt on what was taken from us. We survived Wounded Knee, a massacre. We are survivors. We are fighters. And we are protectors of our land. . . . You mention nothing about the water. You don't want to infringe on native lands, but our water is our single last property that we have for our people. And water is life: Mni Wiconi."[3]

"As history was being told," write Nick Estes (Kul Wicasa, Lower Brule Sioux) and Jaskiran Dhillon, describing this moment in the meeting, "history was unfolding." Nowhere was that significance more apparent than in Young's prescient final words: "I will never submit to any pipeline to go through my homeland. Mitakuye Oyasin."[4] Two years later, Phyllis Young and others from Standing Rock were joined by people from more than three hundred Indigenous nations and many non-Indigenous allies, who gathered to protect Mni Sose and defend the sovereignty of the Oceti Sakowin. Living in camps for nearly a year, these water protectors were pepper sprayed, drenched with water in freezing temperatures, shot with rubber bullets, and attacked by dogs, and some faced prosecution and jail time.

The Lakota phrase that Young used in her testimony, "Mni Wiconi," and its common translation, "water is life," came to play a significant rhetorical role in the actions against the DAPL. Chanted and tweeted by Standing Rock youth in the earliest actions against DAPL, spoken reverently throughout the camps along Mni Sose, found on signs at the many solidarity rallies across the globe, and emblazoned on T-shirts available for purchase online: in public discourse, "Mni Wiconi" and "Water is life" became nearly synonymous with "Standing Rock," which itself has become something of a metonym for the actions in 2016 to protect Mni Sose.

As a rhetorician, I am interested in the power of language and symbols: how they emerge, evolve, and travel across space and time; how they stick to other words and symbols; how they move people and gather them together. The rhetoric of Mni Wiconi first appeared to be an exemplar of the vital advocacy and bioidentification I examine in these pages, and an account of how life might be understood beyond biology, beyond the organism. But as I began working on this chapter, it became clear that bioidentification simply did not fit this case, for reasons I explain below. Rather than abandon the case study, however, I realized that the misfit between theory and case, and the limits it revealed, was both illustrative and necessary.

As I am a settler without knowledge of the Lakota or Sioux languages, there are limits to what I can and should say about the phrase "Mni Wiconi" and the cosmology in which it finds meaning. When such conceptual work is necessary, I turn to the work of scholars from Indigenous studies. But these limitations are more than just ethical, linguistic, or methodological; they are also epistemological. In Western scholarship, there is a tendency to create knowledge from what Santiago Castro-Gómez calls "la hybris del punto cero" (the hubris of the zero point).[5] Walter Mignolo explains that zero-point epistemology "is the ultimate grounding of knowledge, which paradoxically is ungrounded, or grounded neither in geo-historical location nor in bio-graphical configurations of the bodies. . . . Its imperiality consists precisely in hiding its locality, its geo-historical body location, and in assuming to be universal and thus managing the universality to which everyone has to submit."[6] The hubris of the zero point has led many white and Western scholars to present our knowledge as though it was always applicable to everyone everywhere, boldly claiming to speak of *the* world or humanity *as such*. Despite an affinity for the particular and the contingent, even a discipline like rhetorical studies tends to elide its "geographic and embodied location through a universalizing gesture, eliding the heterogeneity against which it functions as a response," writes Darrel Wanzer-Serrano. The resulting homogeneity is a "real, material thing," he explains, but it is also a "thoroughly rhetorical invention."[7] Joining other advocates of decoloniality, Wanzer-Serrano urges rhetoricians to "better situate knowledge in its geographic and embodied specificity and resist attempts to universalize any particular episteme."[8] As Tiara R. Na'puti (Chamoru) argues, centering Indigeneity in particular can help rhetoricians to identify what our field "has inherited from academic predecessors and to uncover colonialism's continuing impacts on our field."[9]

Following Wanzer-Serrano's call, this chapter has two related objectives. The first is to define this book's primary analytic, bioidentification, by situating its epistemic location and marking the edges of its explanatory reach. I do this by excavating the two concepts it joins: *life* and *identification*. In rhetorical studies, there is perhaps no scholar more in need of such work than Kenneth Burke, because of his influence, of course, but also because of his penchant for writing from the zero point. Burke's universalism is most obviously illustrated by the title of the essay "Definition of Man" but is also apparent in many other works, like *Grammar of Motives*, in which he aspires to "lay claim to a universal validity," writes Barbara Biesecker, an objective that "demands that differences be read as surface and, indeed, inessential variations of an underlying structure." Indeed, Biesecker notes, because he is often interested in the "absolute or logically anterior ground of human acts per se, Burke is obliged to bracket out the historically contingent or culturally specific."[10] Likewise, few terms have been more universally applied, and used for more universal appeal, than "life." Yet "life," too, means different things at different times in different places and for different people, and not just because of the various words that people use to name it. Life's slippery referend also shape-shifts across the many times, places, and bodies in which it is invoked.

In what follows, I first show how identification and then the idea of life itself have taken shape in relation to individualism, or what Élie Halévy has called the West's "true philosophy."[11] The centrality of the individual to Western institutions as disparate as Roman law, Christian theology, Enlightenment philosophy, and capitalism in its various iterations is precisely why Frantz Fanon named individualism "first among" the Western values that colonized people must reject in their struggle for liberation.[12] Yet, as I explain in what follows, the individualism at the center of Burke's identification is of a very particular sort, a biological individualism rooted in an understanding of the organism as a distinct entity.

I then turn to the "bio" of "bioidentification," sketching out two approaches to life that have enjoyed prominence in Western science and philosophy for centuries: *vitalism* and *mechanism*. Intentionally reducing these approaches to their most basic meaning, I show how they relate to what Kim TallBear (Dakota) describes as the "Western ontological binary of life/not-life" and what Mel Chen calls the "animacy hierarchy" in which life is ascribed value. I then shift to a discussion of why this location matters and, using the example of "Mni Wiconi," I offer an anti–case study of bioidentification. Case studies typically illustrate a particular theory, principle, or thesis. Here, I explain why bioidentification would fail as a lens with which to view "Mni Wiconi"—not because the Lakota language is arhetorical or because its concepts are inherently Other or mysterious or because concepts that arise in one location are never useful in another. Rather, it fails because of a conceptual misfit. In rhetorical criticism, misfits between theories and texts often have been used to render negative judgment on particular texts or used as a rationale for ignoring them, rather than an opportunity to reexamine the structural mechanisms that enable judgment in the first place.[13] Misfits, in other words, have much to teach us. As Rosemarie Garland-Thomson argues, the problem of the misfit is not found within a square peg, say, or a round hole but in their "juxtaposition, the awkward attempt to fit them together."[14] As the philosopher Brian Burkhart (Cherokee) argues, "marking off the areas of ill-fittedness" is key to understanding the production of knowledge, as well as the relations of power between Indigenous and Western ways of knowing.[15]

Some readers might wonder why I've placed this chapter first instead of last. In much academic writing, and in the humanities in particular, the limitations of our work—if they are mentioned at all—are neatly tucked at the end, almost as an apology that the present study could not explain everything, a failure to reach (or even aspire to) the zero point. But this is an impoverished way to think about limits. Kristin Arola argues that to name the epistemic locations and intellectual traditions you draw from is "to position yourself honestly."[16] Locating knowledge entails limits by definition. This is perhaps especially important for those of us accustomed to speaking effortlessly from the zero point; our task, Christa Olson suggests, is to "learn to sit with limitedness and do so without equating it with apathy or futility."[17] To do this conceptual work up front, and in detail, is an attempt to make scholarship more transparent and more honest—but also more useful. You can examine the branch you are sitting on without sawing it off, after all. Some of us really need better ways of thinking about trees.

Identification

In *Rhetoric of Motives* (1950), Kenneth Burke laid the second plank in a planned trilogy of books that would come to define modern rhetorical theory in the West for many years. Burke's *Rhetoric* positions itself astride the European rhetorical tradition, providing a sweeping view of rhetorical theories from ancient Greece, Rome, and early modern Europe that he uses to establish rhetoric's "basic function" as persuasion. One of the book's most lasting contributions to the field is a theory of the underlying conditions that make persuasion possible: the rhetorical act that Burke calls "identification."[18]

As I noted in the introduction, another word that Burke uses for identification is "consubstantiation," which speaks to identification as a form of rhetorical connection, but one that stops short of total union. Burke explains that "in being identified with B, A is 'substantially one' with a person other than himself. Yet at the same time he remains unique, an individual locus of motives. Thus he is both joined and separate, at once a distinct substance and consubstantial with another."[19] Not only does Burke's theory of identification entail division, but, and this is key, it assumes an a priori division that makes rhetoric both possible and necessary. Identification, he explains, is "compensatory to division. If men were not apart from one another, there would be no need for the rhetorician to proclaim their unity."[20] For Burke, "everything begins with an 'individual,'" writes Diane Davis, which is perhaps the ultimate form that division can take.[21]

In making an early case for feminist rhetorical theory, Sonja Foss and Cindy Griffin take Burke's individualism as their point of intervention. Using the writing of the ecofeminist and neopagan writer Starhawk, Foss and Griffin argue for an approach to rhetorical theory grounded in connection rather than division. In Starhawk, Foss and Griffin find a critical theory of power that reveals competing rhetorics of "domination" and "inherent value." The rhetoric of inherent value emerges from a "life-loving culture" that extends beyond the human world. This rhetoric, they write, recognizes "the inherent value of each person and of the plant, animal, and elemental life that makes up the earth's living body." A rhetoric of inherent value is characterized not in its capacity to persuade or move other humans but by affirming the value of all life on Earth. As such, Foss and Griffin argue, the scope of this rhetoric is not limited to humans but includes "communication among all life forms, human or otherwise, in the earth and the larger cosmos." Whereas Burke is invested in distinguishing and elevating human action from nonhuman motion, which includes the "animal, biological aspect of human being," Starhawk's rhetorical theory rejects the division between human and nonhuman worlds, which are knit together through a shared "life force." This leads Foss and Griffin to a radical reframing of identification: "Instead of experiencing division through each individual's separate body, Starhawk suggests that humans are interconnected with other humans. . . . Furthermore, life forms in Starhawk's theory, are not *identified* with the Goddess but rather *constitute* the Goddess in various forms. Burke's notion of division, which creates the drive toward identification, does not exist in the rhetorical theory generated by Starhawk's perspective. For her, identification or interconnection is the starting point—the initial and primary condition of life—not a desired end point, to be reached through rhetorical activity."[22] Burke's theory of identification is only possible, in other words, and only necessary when one understands the original human condition to be separation, division, and individuation, or what Amaya Querejazu describes as the "modern Western myth" of the universe.[23] In contrast, Starhawk's ecofeminist rhetoric—indebted to the Indigenous thought to which ecofeminism is also indebted—emerges from an understanding of the human condition as connection: connection with other humans, to be sure, but also with the other-than-human world. For Starhawk, this world includes the biotic (humans, plants, animals) but also the abiotic (water, rocks, elements) that are joined together as/by "earth's living body." Here, "life" extends beyond the organism as a shared, immanent force, an ecological and thus rhetorical *and* ethical relationship. "Ethics," Foss and Griffin explain, "consists of an inner sense that each act brings about consequences for everyone since all beings are linked in the same social fabric."[24] In this framework, rhetors are accountable not just for their rhetoric but also for the effects it has in and on the world.

But I want to dig even deeper into this ground. It is striking that for Burke, the origin of human division is nothing short of the act of birth, which, in separating an infant from its mother, gives rise to the central nervous system that governs individual bodies and their subjective experiences, and the "biological estrangement" he locates at the heart of "the" human condition.[25] The Burkean individual, in other words, is a *biological* individual.[26] "I locate the *individual* (as distinct from the kind of 'ideological' identity that is intended in a social term, such as 'individualism') in the human body," Burke writes, "the 'original economic plant' distinct from all others owing to the divisive centrality of each body's particular nervous system."[27] "Estrangement is a biological, indeed ontological, fact" for Burke, writes Biesecker. "[It] is inscribed in the nature of the human being proper."[28] Even further, this biological estrangement is "ontology's insurance premium for securing his entire rhetoric of relationality," Davis argues. "The division between self and other is the 'state of nature' that is identification's motivating force. Identification's job is to transcend this natural state of division, and rhetoric's job is identification."[29]

The concept of biological individuality has had tremendous impact in the sciences, most notably in the Darwinian view of biological community, which "regarded aggregates of individuals of common ancestry as identifiable units in competition with one another."[30] Like individualism more broadly, biological individuality is associated with Western ways of thinking about how we live together. It is fundamentally linked to what Querejazu describes as the West's founding myth—the separation between the human and the other-than-human—which, when joined with liberalism's emphasis on the individual, creates a picture of the human as "superior, distanced, disassociated—separated—from his environment."[31] It is no accident that the concept of the biological individual appeared in Western modernity alongside the citizen-subject at the heart of liberalism, and there are many parallels between these two characters, independence and autonomy chief among them.[32]

But while the biological individuality that Burke takes as fact has indeed been a central concept in the life sciences for many years, there is oddly no consensus as to what it means.[33] Biological individuality may entail "boundedness, integration, the nature of interactions among parts and wholes, agency or governance of parts, propagation by a variety of means, continuity over time, comprising or being part of a biological hierarchy, being a potential unit of selection, contributing to theoretical evolutionary fitness, and, yes, identity or autonomy."[34] Biological individuality can refer to identity—that is, uniqueness, difference, and a sense of spatial and temporal boundaries, which is how it appears in Burke—but also to unity given meaning by scale, as Joshua DiCaglio argues: the unification of cells into organs, organs into systems, systems into organisms, organisms into species, species into ecosystems, ecosystems into the biosphere.[35] The very concept of the organism gives rise to complex questions of self and other, about where one body ends and another begins, putting the entire idea of the biological individual in question. What is the relationship of the cell to the body from which it is taken?[36] Are genetically identical male bee drones individuals? What about the sprawling clonal colonies of quaking aspens? And what are we to make of the symbiotic relationship between the organism and its microbiota—an interdependence so subtle that it renders talk of "the" organism almost meaningless?[37]

In an article in the *Quarterly Journal of Biology*, the biologists Scott Gilbert and Jan Sapp and the philosopher Alfred Tauber argue that the principle of symbiosis inaugurates a paradigm shift in scientific understanding of the biological individual and what it means to literally live together. "We are all lichens," they declare; biological individuals, whether plant or animal, are not a fact but a fiction that has long misguided scientific thinking about life. Even further, grappling fully with symbiosis has the potential to lead biologists in "directions that transcend the self/nonself, subject/object dichotomies that have characterized Western thought."[38] As many scholars have argued, moving symbiosis from an aspect of life to its defining feature heralds a radical departure in the scientific understanding of life, entailing a shift from the discrete individual to the networks, systems, and relations that constitute the phenomenon of being alive. Symbiosis challenges the dichotomies that characterize Western thinking about life as well as the ethical relationships that arise from acknowledging the great many ways that we live together. But there is one dichotomy that remains largely untouched in the wake of the symbiotic revolution, a dichotomy central not only to Western understanding of what life is but also to why it matters: the dichotomy between life and not-life.

The Souls of Stones

As I discussed in the introduction, life is a famously difficult thing to define, and the effort often provokes frustration among scientists, many of whom see the struggles over definition, particularly as it relates to entities seen as natural kinds, as a philosophical issue not relevant to their work.[39] But despite the difficulty, perhaps impossibility, of the task, a great many scientists and philosophers have tried their hands at the grail question.[40] There are not just hundreds of definitions of life; there may be hundreds of taxonomies of life.[41] For the sake of scope, then, I focus in this section on two of the most well-known approaches to life in the history of Western thought: vitalism and mechanism.

The simplest definition of vitalism is that it views life as irreducible to matter. Life, in this view, is something more than something physical or chemical, whether that something is described as a force, a power, a principle, a soul, or Bergson's *élan vital*. For vitalists, life itself is more than an idea or a concept: it is a thing, *the* thing, that makes living entities fundamentally distinct from nonliving entities. "In living things," writes Hans Jonas, summing up this idea, "nature springs an ontological surprise in which the world-accident of terrestrial conditions brings to light an entirely new possibility of being."[42] For vitalists, life is thus not fully explicable by physics or chemistry, which, because they lack a method to account for life's "ontological surprise," are by definition incomplete. The French vitalist Xavier Bichat described this gap as critical to disciplinary self-understanding: "Physics and chemistry are conjoined because the same laws preside over their phenomena. But an immense interval separates them from the science of organized bodies, because an enormous difference exists between their laws and those of life. To say that physiology is the physics of animals is to give an extremely inexact idea of it; I would as much like to say that astronomy is the physiology of the stars."[43] Bichat was writing in the eighteenth century, well before the molecular biology revolution that two centuries later would address the lacunae that life presented to the sciences. But his point here still resonates: for vitalists, there is something *different* about living things that demands a fundamentally different mode of understanding. In "seeking to reduce the specificity of the living," writes Georges Canguilhem, physics and chemistry "did no more than remain faithful to their underlying intention, which is to determine the laws between objects, valid without any reference to an absolute, central point of reference."[44]

For mechanists, intellectually descended from Descartes's famous description of the organism as machine, the living differs from the nonliving merely with regard to organization and complexity. Complicated though living things might be, the mechanist still views them as subject to (and thus explicable by) the same natural laws that apply to everything else in the known universe. One can easily see how the lines of "Science" can get drawn here, as vitalism became not just incompatible with mechanism but derided as an atavistic remnant of a prescientific worldview, rejected and ridiculed as "vapid and vacuous," "downright daft," and an "odious" and "thoroughly discredited" idea wielded as an "epithet" against those who are naïve enough to think that life is something special.[45]

I need to acknowledge the inevitable groans of historians of science here, for I have not only criminally simplified these two views but also reduced a very complicated set of perspectives over a very long time into *two views*, and binary, oppositional views at that. A more thorough account would chart not only how mechanism and vitalism have taken shape as a spectrum through the history of ideas but also the many other perspectives that come to matter in that history—idealism and materialism, say, or more recent approaches to life such as complexity, organicism, or emergence.[46] But by distilling these two views, I hope to bring into focus the dominant assumption they exemplify in the Western understanding of life: a dichotomy between the living and the not-living, the active and the inert, the animate and the inanimate.

In Western thinking, this relationship has roots that stretch back to Greek antiquity. In *De Anima*, for example, Aristotle distinguished living things from nonliving things by their purposiveness. Reason, perception, nourishment, growth, and even decay all indicated the presence of a soul, which for Aristotle defined a living thing.[47] For this reason, Aristotle admits that "even plants, all of them, seem to be alive."[48] As Jeffrey Nealon points out, the plant soul has long been used to mark the "limit of life itself."[49] Stones, in contrast, were explicitly excluded from Aristotelian entelechy, a dismissal that Mel Chen observes "anticipates the affective economies of current Western ontologies that are dominant, in which stones might as well be nothing."[50]

To locate the dichotomy of life/not-life as *Western* might best be seen in contrast with what the science and technology studies (STS) scholar Kim TallBear calls an "indigenous metaphysic."[51] Fleshing out this idea, TallBear draws on the Dakota writer Charles Eastman, who explains in *The Soul of the Indian* (1902), "We believed that the spirit pervades all creation and that every creature possesses a soul in some degree, though not necessarily a soul conscious of itself. The tree, the waterfall, the grizzly bear, each is an embodied Force, and as such an object of reverence."[52] While Eastman's account is obviously not a mechanistic perspective, neither is it a vitalist perspective, for "embodied Force" is found in bears *and* waterfalls, that is, in entities that would not be classified as organisms. For Eastman, each is an ensouled creature by virtue of being a part of creation. What TallBear and Eastman are describing is a world and a worldview in which "life" extends beyond the notion of the biological: "I am comfortable using the term 'life' in ways that go beyond biological meaning as part of my understanding of animacy," TallBear comments.[53]

TallBear illustrates this idea using the very entity that Aristotle discarded from his taxonomy: stone, and specifically pipestone, a soft red stone found in southwestern Minnesota that the Dakota people have quarried by hand and carved into pipes for generations. Life "inheres" in this stone, TallBear explains; "we can describe pipestone as vibrant because without it prayers would be grounded, human social relations impaired, and everyday lives of quarriers and carvers depleted of the meaning they derive from working with the stone." The stone is not just a way of relating; its action extends far beyond what STS scholars might attribute to its status as an "actant." "From a Dakota standpoint," TallBear writes, "the pipestone narrative is one of renewed peoplehood. A flood story tells of the death of a people and the pooling of their blood at this site, thus resulting in the stone's red color and its description as sacred." This is why, she explains, the stone is sometimes spoken of as a relative.[54]

TallBear's description of pipestone is a rich material-semiotic, biotic-abiotic fabric in which "relations" and "life" both have meaning beyond biology. These terms are not tropes (in the way that I might describe the "life" of my computer battery), nor are they "mere" myths to be measured (and by definition found deficient) against a universalized Western understanding of how things *really are*. Rather, Western and the Indigenous approaches to life emerge from "differentiated frames," to use a term from Vanessa Watts (Mohawk and Anishinaabe Bear Clan).[55] TallBear explains that the Western frame, in which life/not-life are seen not just in binary terms but in hierarchical relation, has considerable consequences for humans as well. While scholars of new materialism in its various forms have recently taken an interest in the agency of nonhumans, she argues, their lack of engagement with Indigenous thought, or references to Indigenous ways of knowing as "'beliefs' or artifacts of a waning time to be studied but not interacted with as truths about a living world," denies not just the vibrancy of the nonhuman world but the vibrancy of Indigenous people as well: "Seeing us as fully alive is key to seeing the aliveness of the decimated lands, waters, and other nonhuman communities on these continents. Understanding genocide in the Americas, for example, requires an understanding of the entangled genocide of humans and nonhumans here. Indigenous people cohere as peoples in relation to very specific places and nonhuman communities. Their/our decimation goes hand in hand."[56] What TallBear brings into focus is how "life" within an Indigenous metaphysic means more than a collection of traits or features that living things have in common, more than the "ontological surprise" that makes biological entities special, more than an emergent quality of organisms, more than an ineffable substance or intangible force. TallBear's description of the vibrancy of pipestone and the vibrancy of Indigenous people offers an approach to life that cannot be understood outside of the relations it constitutes—and by which it is constituted.

Burkhart adds another layer onto this perspective. He argues that within an Indigenous focus on locality, which refers to "the manner in which being, meaning, and knowing are rooted in the land," life emerges as "the manifestation of interconnectedness and multiplicity." As such, he explains that "being alive is not dependent on any particular property that a thing might have but on having relationality or interconnectedness itself. Life is not the possession of consciousness, the ability to experience pleasure or pain, the power of self-movement, or any biological process inherent in a particular organism, from this perspective of life. Life is fundamentally the capacity for kinship."[57]

I am not suggesting that the relational meaning of "life" that TallBear and Burkhart describe is somehow its one, true meaning. To do so is to make the colonial move at the heart of Burkhart's critique: "laminating" delocality onto locality, as he puts it. To describe the Dakota understanding of pipestone as *particular* and the Western understanding of the biological individual as *universal* is likewise a misstep: an example of how the colonial hubris of the zero point, and its search for timeless, placeless, bodiless insights, also produces occlusions. If the history of the concept of life in Western thinking—from vitalism to mechanism and back again, from organisms to superorganisms to symbiosis—

teaches us anything, it is that the Western understanding of "life" and its slippery referend has *always* been partial and contingent, has *always* been the achievement of argument, has *always* taken form in relation to its exclusions, and has *always* been particular to the time, place, and structures of power/knowledge that give it meaning.

If life, then, is understood in this frame as the capacity for kinship, applied to biological and nonbiological entities as well as the relations among them, wouldn't bioidentification be *precisely* the name for what is at work in "Mni Wiconi"? In response, I hope by now it is clear why it is so critical to locate rhetorical identification within a Western frame. To understand the notion of identification is to see how rooted it is within a Western notion of identity that is given meaning by the human biological individual. TallBear and Burkhart both challenge the very idea of the biological individual, of course, but more importantly, they also challenge the life/not-life dichotomy in which such an entity has meaning and is accorded value. This challenge complements but also expands Foss and Griffin's ecofeminist-inspired rhetorical theory in important ways. Foss and Griffin note that while Starhawk's rhetorical theory is promising in some respects, it is limited in others. For example, they note that while her rhetorical processes apply to all entities (both living and nonliving), she does not describe "the processes that characterize the rhetorical exchanges between human beings and rocks, for example, or between rocks and trees. The implications of her theory for human communicators and for humans communicating with . . . rocks are left unexplored."[58]

Rocks are always presented as the most difficult case, aren't they? But only within a Western frame, only within that founding ontological dichotomy of life/not-life and the capacities and value it authorizes.[59] Within the Indigenous metaphysic that TallBear and Burkhart describe, it's not difficult to understand how stones might function in communication as more than a medium or subject. I can easily imagine, no metaphor necessary, the carver's hands persuading the red stone, accommodating its softness and hard edges, coaxing it into the pipe it will be, and the prayers it will offer, a process that joins past to present and present to future. In a study of the pipe as a rhetorical object of the Lakota people, for example, David Grant suggests that it "is a potent locus for communication even if it is not itself the medium of that communication or the means by which communication is signified. As a central object for Lakota practices, it is understood as that which can connect Lakota people to the nonhumans and the ultimate great mystery, Wakan Tanka."[60] The action between Dakota and Lakota quarriers, carvers, pipe users, and pipestone is a form of communication, but as Grant demonstrates, it is also a practice that combines material, symbolic, and spiritual realms together in ways beyond Western accounts of the world and the role of language within it.

But appreciating the complex rhetoricity of pipestone is different from, say, describing the relationship between pipestone and the Dakota people as a form of constitutive rhetoric, Maurice Charland's influential rhetorical theory of how collective subjects come into being. Charland argues, building from Kenneth Burke, Michael Calvin McGee, and Louis Althusser, that "'peoples' in general, exist only through an ideological discourse that constitutes them."[61] While Charland's "people" is a useful tool for understanding the rhetorical dynamics at play in the formation of many collectives, including nations, Christa Olson argues that it "is primarily constituted through explicit acts of identity-forming speech rather than viewed in terms of a broader constitutional scene, leaving aside the myriad prevalent yet implicit elements that promote membership and identification."[62] What place, we might ask, is there for pipestone in Charland's "general" theory of peoplehood, what role is there for the turtle (which plays a key role in the Anishinaabeg account of creation), and what role is there for land in the constitution of so many Indigenous peoples around the world? Once again, it's not that rhetoric has no place in these stories of creation or in the constitution of Indigenous peoples in the present. But to suggest that the Dakota people "exist *only* through an ideological discourse," as Charland puts it, or to flatten pipestone into a trope or symbol presses a colonial frame onto the story, an occluded Western interpretation of the lively relations among human and other-than-human worlds in which pipestone participates. Likewise, to speak of the Dakota people "identifying" with the stone is a misfit, and for the same reason that "bioidentification" does not fit "Mni Wiconi." To speak of "blood" as a common substance that rhetorically joins the Dakota people to the red stone or "life" as that which rhetorically joins the Oceti Sakowin to Mni Sose *is to see them as separate to begin with.*

More Than a Slogan

"Mni Wiconi" and "Water is life" are often described as the "slogans" of the actions against the DAPL. But what does it mean to call these phrases *slogans?* A slogan is a short, memorable, "social symbol" that evokes an emotional response and functions to unify people, often "with widely varying motives and beliefs to focus their protest on a common target." A key function of a slogan is to condense a complicated message or ideology into something simple. Slogans thus function as synecdoche; consider, for example, Barack Obama's 2008 campaign slogan, "Yes, We Can!," his version of the United Farm Workers' slogan "sí, se puede," which stood in for his campaign's hopeful focus on the future.[63] Slogans are meant to be remembered; they are evaluated in terms of their *catchiness*; they are meant to *circulate*. This is all the more so in the digital age, when slogans often appear as hashtags on social media, where success is measured by spread. Slogans are a rhetorical convenience. They save time and effort. It's no surprise, then, that the term "slogan" has taken on a negative cast. "Slogan" has become synonymous with empty words that circulate widely but lack action, perhaps the merest of "mere rhetoric."

There is no question that both "Mni Wiconi" and "Water is life" *functioned* as slogans in public discourse about the 2016 actions at Standing Rock. Nicole Metildi, for example, has suggested that "Water is life" might be seen as a coalition-building rhetorical ecology that drew together the defense of Mni Sose, the movement for Indigenous rights, and movements for environmental and social justice. To illustrate this point, Metildi uses the words of Caro Gonzalez, interviewed at the height of the action at Standing Rock, who explains, "So many people are dealing with immense poverty and racism. This personal hurt we feel, that's so hard to voice. Like it's so much easier to say 'mni wiconi,' water is life, than to say 'this is my life and this is what's happening to me.' And through that, through mni wiconi, we've actually been able to talk about missing and murdered Indigenous women. We've been able to talk about misogyny and colonial mentality. You know I never really thought that an ecological issue would be something that ties everything together."[64] In this description, "Mni Wiconi" exemplifies the slogan's unifying function ("ties everything together"): it brings people together but also ideas, represented in phrases like "Stand with Standing Rock" and hashtags like "#DecolonizeWater." Gonzalez explains that it's "so much easier to say 'mni wiconi'" than to articulate the larger story, more difficult to voice, of experiences of racism, misogyny, poverty, and the persistent violence of colonialism. If we see "Mni Wiconi" as a slogan, this might be an exemplar of its function as synecdoche.

But to see "Mni Wiconi" *as* a slogan selects a particular reality, which deflects another reality by default.[65] To begin to appreciate what it is deflected, it is necessary to understand "Mni Wiconi" in relation to the Indigenous frame that

TallBear and Burkhart described earlier and especially in relation to Lakota cosmology, in which the meaning of "life" is tied to the capacity for kinship, in which language and knowledge are inextricable from the land. "In the semantics of delocality," Burkhart explains, "words and meaning are abstracted from their originating context in the land, which always starts with here and now and in the dirt in front of me. . . . Language that is delocalized expresses another *ego conquiro* through the idea that a word can carry and even force meaning and reference beyond locality and across localities."[66] Within this frame, "Mni Wiconi" is directly tied to the place from which it emerges and the fabric of concepts and entities from which it acquires meaning. Estes and Dhillon provide additional context, which they link with the actions of the Standing Rock Water Protectors: "The Mni Sose, and water in general, is not a thing that is quantifiable according to possessive logics. Mni Sose is a relative: the Mni Oyate, the Water Nation. She is alive. Nothing owns her. Thus, the popular Lakotayapi assertion 'Mni Wiconi': water is life, or, more accurately, water is alive. You do not sell your relative, Water Protectors vow. To be a good relative mandates protecting Mni Oyate from the DAPL's inevitable contamination. This is the practice of Wotakuye (kinship), a recognition of the place-based, decolonial practice of being in relation to the land and water."[67] Note how Estes and Dhillon describe "Mni Wiconi" not as a slogan but as an "assertion," a statement of fact about the relationality of the Oceti Sakowin Oyate and the Mni Oyate. In this way, then, "Mni Wiconi" is not a *synecdoche* for the lives of the Lakota people or their nonhuman relatives or of their fight for sovereignty and survival. A synecdoche uses a *part* to stand in for the whole, which, once again, suggests that they were divisible to begin with.

Burkhart digs even deeper into "Mni Wiconi," in a passage that is well worth quoting at length:

> In the Lakota language, *Wakȟáŋ Tȟáŋka*, which is usually translated as God or Great Spirit, is really a vessel that channels movement or energy (from the word *wakȟáŋ*) and is a part of everything (from word *tȟáŋka*). When the Lakota people say *Wi* is wakȟáŋ (the sun is sacred) or *Hanpewi* is wakȟáŋ (the moon is sacred) they are referencing the more than material sense in which the sun is the continual source of all life on earth. They are referencing the movement or energy that is wakȟáŋ as manifest in the sun and the moon, but also that the movement or energy is also the center of the spiritual strength that allows the *Oceti Sakowin* to stand against the

> overwhelming physical and legal power of the Dakota Access Pipeline as an agent of the settler state. It is the same for the words *mni* (water) and *wiconi*, (health or life continuing) which comes from the root word *ni*, which means life. Dr. Cheryl Crazy Bull describes the intersection of these terms in this way: "An origin story of the *Oceti Sakowin*, tells us that the blood of First Creation, *Inyan*, covers *Unci Maka*, our grandmother earth, and this blood, which is blue is *mni*, water, and *mahpiya*, the sky. *Mni Wiconi*, water is life." A real understanding of this, she continues, gives the *Oceti Sakowin Oyate* (the people of the Seven Council Fires) "the tools they need to indeed manage all aspects of their lands, which is our life blood, and the source of our lives" (2016). Water is indeed life, then, and in a more than material sense. *Mni* is the life-blood of the earth and the sky, which is why the word "*mni*" (water) has the word "*ni*" (life) in it. It is also the life-blood and spiritual strength of the *Oceti Sakowin Oyate* as well.[68]

Key in Burkhart's explication of these terms is how "wakȟáŋ" and "ni" extend the meaning of life beyond living things but also beyond a vitalist understanding of life as force or energy. The Lakota word for water *contains* the word for life: as Burkhart points out, water is life *indeed*.

"Life Finds a Way"

I now want to turn to a final example of a misfit, the invocation of "Mni Wiconi" by an emergent group that calls itself Vitalist International (VI). While details about the group or its members are difficult to come by, its Twitter account explains that it is "an international network of artists, poets, martial artists, farmers, painters, musicians, lovers, writers, filmmakers, and creative people dedicated to the liberation of the senses from economic and political strangle-hold."[69] With clear influences from situationism, anarchism, communism, and Bergson, VI describes itself as an "autonomous and distributed network" dedicated to "proletarian research into our uncertain present." On the group's Twitter bio, its location is listed only as "Earth." In early 2020, VI published a manifesto in the magazine *Commune* (tagline: "the answers are in the streets").[70] In it, VI explicitly positions itself as building on the actions of the Standing Rock Water Protectors, and "Mni Wiconi" is invoked as the manifesto's source of inspiration. When read with reference to the Indigenous metaphysic described earlier, however, VI's manifesto illustrates what happens when "Mni Wiconi" is understood only as a slogan. Drawing on that key term in Western understanding of life, "vitalism," VI clothes itself in the trappings of relationality that "Mni Wiconi" evokes while also fundamentally misunderstanding what it means.

In a nod to *Jurassic Park*, the VI manifesto is titled "Life Finds a Way." It begins with praise for the actions against the DAPL, which are described as the apex of a long line of revolutionary acts: "Since the era of revolutions opened over two centuries ago, no greater slogan has emerged than the simple dictum uttered in the fight against the Dakota Access Pipeline: *Mni Wiconi*—Water is Life. It would be perfectly natural, then, if the next revolutionary slogan were more direct. In a kind of advanced concision, it would unleash an even greater and less mediated potential: life." A focus on life leads to a vitalist worldview, which is described as "radiant intuition," a "coordination of the human body with bodies of thought, bodies of water, with bodies of buffalo charging at police, of life forms with art forms." The vitalism of VI has a direct lineage to Bergson's *élan vital*, in its description of life as a "creative force" that might be found "in the woods, at punk shows, at the beach, in dance parties, in the black bloc, wherever screens do not loom so large." VI celebrates the aesthetic and the sensuous while condemning "the great structures of power—race, gender, private property, the state"—which are described in mechanistic terms as "guarantors of separation," in which "life assumes the form of the machine." In contrast, the manifesto affirms, "vitalism names the resonance between the rivers, the trees and the forests as much as between objects, spirits, the dead and those banished to social death. The depravity that is the desert and the warmth that is the sun we take as simple evidence of the fact that in life, we are not alone."[71] *In life, we are not alone*—this phrasing, as well as descriptions of the human body "coordinating" with bodies of water and thought and buffalo, connections between biotic and abiotic entities, and terms like "resonance" call to mind the Indigenous metaphysic described earlier, but a closer look reveals several critical differences.

First, unlike Estes and Dhillon's description of "Mni Wiconi" as an "assertion," VI describes the phrase as a "simple dictum" and a "slogan." Despite the purported greatness of the "slogan," however, VI calls for a more "concise" revision. If a slogan is a rhetorical convenience, VI calls for an even more slogany slogan, as it were, greater rhetorical efficiency, stripping away the "water" of "Mni Wiconi" to focus *only* on life. But strikingly, the concision of "life" is also described as more "advanced," which immediately implies antonyms with long, violent histories: it suggests that "Mni Wiconi" is *backward* and *primitive*.

It's clear that in the manifesto, "life" here means *biological* life. Only with a biological meaning of life in hand, and only when life is positioned in a hierarchical relation of value with not-life, can one dare speak of the "depravity" of the desert, and only then if one has no knowledge of the delicate ecosystems and the many lives and ways of living that the desert makes possible. VI evokes a cartoon of a desert, an empty, "barren" space that also evokes the *terra nullius* arguments that European colonists used to justify their violent settlement of the Americas.[72] This representation of the desert is all the more striking when VI describes its project, without explanation, as "somewhere between the great Apache warrior Lozen and Audre Lorde."[73] It's worth noting that Lozen, a Chihenne Chiricahua Apache woman, was a famed warrior who fought US forces alongside Geronimo and who once called the "depraved" southwestern desert lands her home.

To see "Mni Wiconi" *only* as a slogan is to assume that the words that compose it not only are translatable without remainder but also can be unmoored from their "originating context," or what Burkhart calls the "semantics of delocality." I am reminded that the etymology of "concision" is to *cut off*; by cutting water out of life, the manifesto also cuts life off from the relations it constitutes and by which it is constituted. But most troublingly, the Oceti Sakowin are also cut out of the manifesto entirely—both as subjects and objects of the struggle against the settler state and corporate power—in favor of an amorphous, agentless, general "fight against the Dakota Pipeline," which the manifesto suggests is undertaken *on behalf of life itself*.

Conclusion

TallBear insists that seeing Indigenous people as "fully alive" is "key to seeing the aliveness of the decimated lands, waters, and other nonhuman communities on these continents." As Burkhart observes, "Mní" names the enduring strength of the Oceti Sakowin, a strength that enabled them to survive the apocalypse of European colonialism and that drives them to resist threats to their land and sovereignty that continue to this day. This is the context in which the people of Standing Rock and their Indigenous and non-Indigenous allies describe themselves as "protectors" rather than "protestors," a role that Estes and Dhillon explain is "mandate[d]" by "being a good relative."[74] When your relative is under attack, you don't *protest*; you defend, you protect. Interviewed at the height of the DAPL actions, the Mdewakanton Dakota and Diné activist Dallas Goldtooth explained that this terminology is more than a matter of semantics: "The term 'protector' speaks more true to why we are here and what we are doing. . . . The struggle that you see here is not built out of hate but built out of love. It's not really what we're fighting against, it's what we're fighting for."[75]

To see the struggle for land and sovereignty *as* love is to appreciate the many visible and invisible connections that create and sustain communities over time; to see stone *as* life or life *as* water is to appreciate the complex relations among human and nonhuman, biotic and abiotic worlds that cannot be disentangled. To see "Mni Wiconi" *as* part of an Indigenous metaphysic is to appreciate how wrongheaded it is to think of it as a slogan or as a form of rhetorical identification. Both are rooted in an understanding of life and language that presumes that humans can be separated from other forms of life, that life itself can be separated from the process of living, that life is separate from not-life, and that knowledge and language can be separated from the land. To see life *as a capacity for kinship*, as Burkhart puts it, is to begin to appreciate the many threads that weave through a particular place over a time scale that is difficult for settlers to fathom, relations of mutual responsibility that cannot be accounted for by Western individualism and the ethical and political systems it gives rise to. In the interview that follows, Kyle Whyte and I discuss this idea, and its implications, in more detail.

Kinship, Consent, and Mutual Responsibility | A Conversation with Kyle Whyte

Kyle Whyte is George Willis Pack Professor of Environment and Sustainability at the University of Michigan. His research addresses environmental justice, focusing on moral and political issues concerning climate policy and Indigenous peoples, the ethics of cooperative relationships between Indigenous peoples and science organizations, and problems of Indigenous justice in public and academic discussions of food sovereignty, environmental justice, and the anthropocene. He is an enrolled member of the Citizen Potawatomi Nation. Kyle currently serves on the White House Environmental Justice Advisory Council, the Management Committee of the Michigan Environmental Justice Coalition, and the Board of Directors of the Pesticide Action Network North America. He has served as an author for the US Global Change Research Program, including authorship on the Fourth National Climate Assessment. He is a former member of the Advisory Committee on Climate Change and Natural Resource Science in the US Department of Interior and of two environmental justice work groups convened by past state governors of Michigan.

JJ: I've really been inspired by your definition of kinship relations as "moral bonds expressed as mutual responsibilities."[1] Could you talk for a bit about what that means?

KW: Sure. One thing to back up on is that, in Indigenous studies, a lot of people have been writing about kinship. And I think there's so much more that we can do, each of us writing about kinship in ways that are similar to each other and writing about it in different ways too. Because the way I understand kinship is that it's a whole set of roles and universes of thinking about morality and moral bonds, relationships that we have with each other that have an ethical significance to them.

And there's a lot of different moral bonds. A right, for example, is a moral bond. An obligation is a moral bond. And there's other relationships as well that either are moral or have a sort of force to them, like a contract, for example. But kinship, for me at least, refers to moral relationships that have a particular purpose.

The purpose, in my view, works something like this. Consider when you're in a kinship relationship that you take seriously, even if you just practice that relationship in your interactions with, say, one other person or one other nonhuman. In that relationship, you know that your participation in that relationship contributes to making it possible for the community that you live in to provide the support network that everybody needs so that they can live freely, so that they can be self-determining. And so, a kinship bond is one where, when I take it seriously, it's not serving a local purpose. It's serving a community purpose.

I understand kinship has having a couple of things at play. For any relationship you have with somebody, human or nonhuman, there's the type of relationship that you're in. So, a contract or a right—those are just types of relationships. A type of relationship that is common in kinship is responsibility or mutual responsibility. And mutual responsibility doesn't necessarily mean that each relative in the relationship, each party in the relationship, owes the other the same actions. They can actually be pretty different, both between humans but also especially between humans and nonhumans.

So, responsibility is just a type of a relationship. So, if I have a responsibility to you, for example, it means that we each have expectations that our well-being is connected to interdependency—that community purpose. We acknowledge each of us are dependent on each other in different ways and that each of us has value and expertise and attributes that can support the well-being of the other. Those expectations mean that we acknowledge we have responsibilities to each other. And the thing about responsibilities is that they're almost never governed by explicit rules.

So, a right, for example, usually has some rule attached to it, or it's expressed by a rule: "I have a right to this, a right to that." But responsibility does not depend on the articulation of a rule. A responsibility is something that we will practice, regardless of what the rules or the law says. It's a type of interdependency that is more intimate to our family settings, our cultural settings, and many other settings in which we realize that the law or a particular policy is not actually the only, or even the best, way to express our morality, our moral bonds to each other.

Now, the thing about kinship, though, is that we can describe the type of relationship as mutual responsibility or one type of kinship. I think there're probably others too. But actually, what makes the mutual responsibilities truly meaningful is if they have certain qualities attached to them. For example, we know by having this conversation that we have mutual responsibility. But if we don't respect each other's consent or reciprocity, if we don't have trust, then it makes it pretty challenging to exercise a mutual responsibility.

So, there's a difference between a *type* of relationship and a *quality* of relationship. I've been in many environments, and obviously, in higher education, this has become a prominent issue in the last three or four years. You have well-meaning people that understand that they have mentorship responsibilities. But, if there's no consent or if there's no reciprocity, then it's meaningless, right?

JJ: Right.

KW: So, just having a responsibility doesn't mean that you actually have kinship. To actually have kinship, it means that your responsibilities are tied to these qualities, and there's a high level of consensuality in the exercise of the responsibility. There's a high level of reciprocity, a high degree of trust. And you can talk about more qualities, too. It's not just those three. Accountability could be considered a quality. Transparency could be considered a quality. But there's also the idea of privacy; not everything needs to be out in the open. There's some types of secrets that actually have a positive role. And people can have a respect for certain types of privacy and nondisclosure.

Now, here's the thing about the qualities, though. They, too, are not based on rules. But, even more so than responsibility, they have a particular embodied character to them. So, for example, for me to take consent seriously, that doesn't mean I just acknowledge that consent is important. We can all do that instantly. But actually, consent is very embodied.

JJ: Interesting. How so?

KW: For example, if I'm in a situation where I'm the privileged party, and I say I respect the consent of somebody else who doesn't have the same privilege, they can say, "Well, wait a minute. I hear you say that, but we're in an institutional situation where I know my consent doesn't matter. My consent hasn't mattered in my lifetime and hasn't for generations for people like me." And, in that way, I would suggest that's a situation in which we're at the very ground level of establishing kinship.

We're trying to get to a situation where consent is embodied. And it's more than just, "I respect your consent," but that everything that I do—intentional, unintentional—reflects consent. And that's where you have a relationship that the exercise, at the embodied level, at the explicit level of consent, can be shown to contribute to a larger culture or society of consent. That's a society that places a premium on consent, that can actually provide that support network for each person, right?

JJ: Right.

KW: So, when times get tough for me, I know that I have a community of consent there to support me. And we can say similar things about reciprocity or trust too. And just one other example with reciprocity: Reciprocity is not just an exchange. It's not just, "I give you a gift. You give me a gift." It's actually deeper than that, because I can give you a gift, but, if we were in a different position with respect to privilege, then that reciprocity doesn't really count as an expression of kinship, right?

For reciprocity to truly be established at the embodied level, it means that, if I give to you, even if you never have a chance to give back, I know that my generosity is not in vain. But, to get to that emotional level with somebody, including with a nonhuman, can't just happen quickly. It's not just, "Well, I know that's what I need to do. I'm going to do it."

To be in a kinship relationship means that the moral bonds around you, at one level, are mutual responsibilities, but they're meaningful and exercisable because they have these qualities that are attached to them and that are practiced as part of the implementation of those responsibilities.

JJ: So, the last thing you said, about how this relationship is not something that happens quickly, that it's a process that unfolds over time, is that because kinship is tied to place? Is that part of where that sense of responsibility comes from? A kind of enduring relationship with a place?

KW: Yeah. I appreciate the question. It's actually a fairly challenging question because, at one level, I think that kinship does emanate from the sense of place that would come from living in a place for, say, a long, long time. But, on the other hand, I think it's worth looking more into what that really means. Because I don't think it's by virtue of, say, having a long history in a place that automatically generates kinship. I think what we're more dealing with today is that the sphere of ethics that kinship is part of, at least in my opinion, is something that is particularly lacking right now. The way that dominant institutions, whether it's private industry or government or education or nonprofits, are treating people today? Their failures are indicative of a lack of kinship.

Circling back to the connection to place, when I think about where kinship comes from, what's important to note is, for some Indigenous people, their histories are not always about being in one single place the whole time. Some Indigenous people's histories are fairly migratory, and they talk about movements and oftentimes pretty large geographic movements, like the Anishinaabe great migration. But the kinship traditions persisted over time. And the kinship traditions changed and transformed over time, if you look at a lot of Indigenous stories, fairly older ones.

Actually, a lot of stories involve problems to the survival of those societies. And then, people experiment with solutions and may fail or may succeed. But the lessons that come out of it are not things like, "We could have avoided this problem if we had a better technology." Usually, the lessons are things like, "Well, if you were more reciprocal with animals or plants, you wouldn't have this problem" or "If you respected the independence of certain plants or animals, then you would have been in better shape."

The lessons that arise are kinship lessons. And I think kinship, because it is tied to informal relationships, has a couple of features to it. So, on the one hand, yes, kinship is related to long-term inhabitation of a place if we think of what it would be like to remember a society where the kinship was really, really good. Imagine if you were thinking about either your family or a team that you've been on or something like that, where people really were about consent and reciprocity and trust, where all those things operated at a high level. And then, you encountered a huge challenge, a huge disruption, and everybody was able to cope effectively and support each other, right? One might recognize that it took many years to get to that point and the people were together in the same place. They shared a space with each other. And that's what created that kinship. But see, at the same time, it's also recognizing that that kinship is what you need to be able to adapt to disruption.

JJ: Sure.

KW: If you're forced to move from a place where you've lived a while to a new place, it's kinship that gets you through that change, right? In terms of the analysis of different types of colonialism or racial capitalism, I think this is where it's important to understand the specific ways in which these forms of oppression and power have affected kinship. It's not just the dispossession of the lands. If you look at a lot of capitalist colonial histories of the wielding of power against Indigenous people and a number of other groups, it wasn't just about land dispossession. Because when, say, the United States forced native people into private property regimes or into smaller reservations, Indigenous people adapted very quickly to that and developed communal kinship-based systems for coping with that change. Then the United States would swiftly come in again and either outlaw or discourage or do something further that would eliminate people's capacity to do that. One of the low points in US policy being the boarding schools, which were among the ultimate expressions of that form of antikinship violence. So, anyways . . .

JJ: That actually raises another question, which is the extent to which this idea of kinship works in fights against colonialism. How do you think that this idea grounds or shows up in Indigenous struggles for sovereignty or justice?

KW: Yeah, absolutely. So, there's a lot of ways to understand justice. But a fairly common one, within European or American, Canadian, those types of traditions, is, when you're talking about justice, you're talking about something at the level of societies, how societies relate to each other and groups relate to each other. And so you can actually have an injustice tied to violations of kinship.

One of the things that you see is that, if you look, say, at the history of treaty-making, a lot of native people really did understand the treaty-making process was actually a kinship process. Kinship is not really tied to a biological family. There are different layers of kinship. You can have kinship that operates in a family situation. You can have kinship that operates in a work or professional situation. You can have kinship that operates at a political level. You can even have kinship that operates at a global level. There's a way in which kinship can be connected to the responsibilities that we have to people we don't know in other parts of the globe.

So, kinship itself need not only be only a local form of ethical behavior. It actually can extend and kind of scale up. It looks different at each scale. So, what happens—and I'm not necessarily saying this is the view, but it's one way I've kind of made sense of colonialism—is that, when a population is subject to colonialism, typically what they're going through is that another society is seeking to undermine the kinship that that society has and the kinship potential of that society.

Colonialism very quickly tries to attack consent and self-determination. It tries to attack reciprocity or genuine forms of reciprocity. It tries to attack trust and to foster distrust. And it does so through mobilizing forces that do this operation on kinship, both between the aggressor society and the resisting society but also within the resisting society, and then creates dissension. And, over time, that makes it harder and harder for a society to function without kinship relationships within it.

What I think is most challenging about kinship, in terms of analyzing power, is that you can look at it a couple different ways. I argue that we can talk about kinship in relation to Indigenous people, people that have suffered violence. We can also talk about kinship with regard to powerful populations that are seeking to destroy kinship relations in other groups.

The way I understand it is that, if you look at, say, the dominant society in the United States, do they have no kinship? Well, no. There is a whole thing about kinship. But you have to understand it a little bit differently than how we're used to talking about it from a resistance or an anticolonial perspective, right? So, for members of the dominant society, one, do they value kinship? Yes, they do. They just don't admit it, because they have these myths of individual success and hard work and so on, when we all know that, without social networks, friendships, connections, networking.... But sometimes they do admit this, right? They'll say, "Well, it's about who you know." But it's always so mixed up in this doctrine of individual success that they downplay actually that element; when they talk about networking or other things, they're actually acknowledging that it's a fabric of kinship that makes possible what they deem to be successes, right?

JJ: Oh, exactly.

KW: So, yes, the kinship is there. Do they struggle explicitly with kinship? Absolutely. One of the largest problems that nobody in the dominant society seems to be able to solve is that they have large institutions, companies, universities, governments, others. They have all of this bureaucratic code. They have rights. They have laws. But people still abuse each other within them, right? So, yeah, you have a right to say, "Equal pay." That's a reciprocity type of thing. But is that how it plays out in real life? No, because there's all these informal ways to destroy that—either by leveraging a lack of kinship or by leveraging, say, the kinship of patriarchal men to each other. So, you can talk about kinship in that way, and it's a very toxic form of kinship, right? Like with other types of ethics, kinship has its pros and cons. I've emphasized a lot of positive qualities, but kinship can very quickly become toxic and can become extremely problematic.

So, we can actually talk about the dominant culture in terms of how kinship operates within it but also the lack of kinship and the concern that people have about what that lack of kinship means for the future. We see that this lack of kinship has been tied to things like not being able to create laws and policies and public institutions that protect people's safety. And that's made it hard to mobilize on things like climate change or other key topics where there's just stagnancy and not a lot of movement.

JJ: Absolutely, yes. I was also thinking about how you can have a biological family without kinship, that you can have plenty of toxic relations there and lack not just mutual responsibility but any responsibility at all. So, I want to return to something you said earlier, about the way that mutual responsibility is also part of human relations with the nonhuman world. I think that you turn to Deborah McGregor on this point in your essay, who writes not just about human responsibility to water but water's responsibility to us. I'm wondering, what role does "life" have in this system of relations? Does it disappear in favor of these other kinds of boundaries, if it is the case that both living and nonliving things in this network have mutual responsibilities?

KW: Sure, absolutely. So, I actually think that, when Indigenous people but also other groups too talk about a politics of life or an ethics of life, that's oftentimes what advocates say. They contrast Indigenous respect for life with capitalist greed for resources. But I actually think there's more to it when they say that. They're using a few words to express what, for them, is a bigger reality. And kinship represents actually a lot of different aspects of ethics, an actual way of thinking about everything around you, right?

So, for example, if you and I both have rights, obligations, to each other, and actually, if that was all we had with each other, that's how we'd see each other. We'd understand. We'd view each other as rights holders or the patient of obligation from a right. So, it's always important to note that, when you believe you're in some kind of relationship, that you're actually seeing the party or the relative in that relationship in a particular light.

Now, of course, we're not just confined to one thing, one type of relationship. There's a ton. And kinship itself, I think, brings with it an actual way of thinking about the world. You can think about the world only through kinship, if you wanted to. And I think kinship makes it possible to understand all of your relationships, all of the different relatives, as living, whether, in a Western scientific sense, they're deemed as living or not.

Kinship opens up a way of relating to anything that makes it living in a way. And I'll say what I mean by this. Take a quality like consent. Consent can be experienced in a lot of different ways, depending on the relative in question that you have a responsibility to, that they have a responsibility back to you. So, in a situation like you and I right now, where we're able to speak English and probably be able to come to some common understandings, I think we can have a type of consent where each of us can be confident in our self-determination and being informed and so on. We can have a clear kind of consent. We can both consent to whatever it is that we're going to work on together or be part of together.

But what does consent look like if it becomes a way of thinking about our relationships with animals, for example, or with insects or fish or plants? Well, in this case, we know that we do things to them. We can take actions to them that can make changes in them. We can kill them. We can do some things that, by our standards, are pretty violent. We could do things that we know have kind of a nourishing or cultivating effect. But we don't actually ever really know what that means to them, right? We never really know. And so, we know that we can take actions that affect them and that their actions affect us, but we're never certain that they consent. And I don't think we will ever be certain because we don't necessarily have that same type of communication.

JJ: Yeah, that seems key.

KW: If you look at a lot of Indigenous people's ceremonies and other activities, they can actually be explained as a constant reminder and acknowledgment that we're dependent. We're in mutual responsibilities with a number of other beings, and we don't know the consent status of our behavior toward them. And it's kind of a never-ending process of being patient, of trying to learn, grappling with the dilemmas of death and harm and ecological change. The minute we believe that we do know that they've consented is, I think, when we've really, really gone too far. And we've asserted our humanness over and against their interest. And so, I think it's important that there are consent relationships out there where you don't know if the other relative has consented. And we struggle with that. And we have to remind ourselves every year, every season, whatever it is, that we're very, very grateful for the fact of that interdependency and respectful of the fact that we might never know what the consent status is.

Then there's another type of consent, which is a little bit different, which is that there are systems or beings or other entities out there that do things to us that *we* don't consent to—you know, a weather event, a storm, climatic conditions, or even different types of insects, right? They do stuff to us that we don't consent to, that we're not always entirely happy about. So, how do we negotiate that situation? It's a recognition that we understand that we don't consent to what they do to us. We also understand that, if we try to assert our will over and against those things, we'll destroy aspects of the ecosystem that would be potentially deadly to us. So, it's a sense of interdependency where there are some beings that we don't consent to.

In different Indigenous traditions, there is a way of thinking about, "Well, wait a minute. When this big storm happened that we didn't expect, did we cause that, or was that caused by something else?" or "How do we get prepared for that? How do we respond? How do we understand the causality of that from the standpoint of responding in the most respectful ways?" And so, in this way, we can actually look at something like consent from all these different perspectives that affect our interactions with the nonhuman world.

In a way, I think that kinship means that we don't need to get caught up in whether there is any definitive answer in the English language about whether some things are living and some things are not living. Instead, we just focus on grappling with how reciprocity, consent, trust, and other qualities operate, depending on the dynamics of those relatives in relation to my own dynamics, my own agency, my own capacities. Engaging with nonhumans through such qualities opens up a number of relationalities.

JJ: That's fantastic. I'm just buzzing with ideas here. It raises so many interesting possibilities for how we think about the relationship between agency and responsibility, but I suppose that's a conversation for another time. Any final thoughts?

KW: Yeah, because of the situation with injustice, a lot of kinship conversations, including a lot of the conversations of kinship in the feminist literature, such as in feminist science studies, we've had to offer kinship as a form of resistance, a response to different types of inequities. It always comes across as very much centered on human-to-human relationships or applying human-to-human relationships to relationships with nonhumans. And I just think there's so much more there. I think the more we kind of experiment with the details of what kinship accounts look like, if we just have the space to talk more freely about it, instead of always having to be in resistance mode, you'd get so many interesting, provocative, and empowering meanings of a really powerful concept like consent. I think that it's important.

2

A Sense of Commonality | Bioidentification in Deep Ecology

There is no thinking of life that is not precarious.
—Judith Butler, *Frames of War*

Deep ecology is an environmental movement that promotes the idea that all living beings have intrinsic worth, and it urges systemic change in our social, economic, cultural, and political structures to ensure the survival and flourishing of the natural world. First articulated as such in 1972, deep ecology has been controversial from its inception and has attracted critics across the ideological spectrum. Some see its talk of self-realization and communion with the other-than-human world as New Age nonsense; some philosophers have suggested that its logical and ethical framework is inconsistent; ecologically minded feminists, anarchists, and Marxists have criticized its thin attention to social and political issues; decolonial scholars have criticized its Western bias and limited engagement with Indigenous thought. These critiques have been so persistent and widespread that some environmental philosophers have suggested that deep ecology has outlived its usefulness (if it was ever useful to begin with). In the well-known textbook *Environmental Philosophy*, for example, the section on deep ecology was recently eliminated from the fourth edition.[1]

Although the scholarly value of deep ecology might be in question, there is no doubt that it has had a tremendous impact on environmental thought and activism, especially in Europe and the United States. The profound feelings of connection it evokes with the other-than-human world can inspire a fierce commitment to political action.[2] Earth First!, for example, perhaps the best-known radical environmental group in the United States, has long had a symbiotic relationship with deep ecology. The founders of Earth First! explicitly based the organization's goals and objectives on deep ecological principles, which have been described as the "mythology" of the movement.[3]

The vast ideological and intellectual scope of deep ecology make it difficult to grasp as an artifact of analysis. "Anyone who attempts to reconcile Heidegger's with Leopold's contributions to deep ecology finds the going rugged," comments Max Oelschlaeger in *The Idea of Wilderness*. Deep ecology, he suggests, is thus better thought of as "a diverse collection of ideas [rather] than a well-defined paradigm."[4] As such, I join Joshua DiCaglio, Kathryn Barlow, and Joseph Johnson in their assertion that deep ecology is best viewed as a *rhetoric*, rather than as a unified philosophy or political program. Deep ecology seeks to create metanoia in its audience, they argue, a "change in mind" that produces an ecological view of oneself in relation with the natural world.[5] Importantly, this metanoia produces not only a shift in the audience's values, beliefs, or political commitments but also a transformation in their sense of identity.

In this chapter, I examine the rhetoric of deep ecology primarily through the work of Arne Næss, the Norwegian philosopher who coined the term "deep ecology" in 1972 and who was the "presiding icon" of the movement until his death in 2009.[6] Identification is central to Næss's approach to deep ecology, and he names it as the "source of deep ecological attitudes" that are key to the metanoia that DiCaglio, Barlow, and Johnson describe.[7] In what follows, I first provide a brief biography of Næss and his development of deep ecology in the early 1970s. I then examine the role that bioidentification plays in deep ecology, Næss's work in particular, as well as critiques that problematize the role of self and other in its call for solidarity with the other-than-human world. Ultimately, I argue that the vital rhetoric in Næss's writing, and in deep ecology discourse more generally, reveals bioidentification as the means by which deep ecology rhetorically transforms living matter to *a* life, that is, an entity of moral significance, a form of what Allison Rowland calls a "zoerhetoric," material and discursive practices by which "existents become intelligible" and are accorded value and protection.[8] To conclude, I briefly examine how deep ecology's vital rhetoric has been taken up by the far right, focusing on the writing of Pentti Linkola, a Finnish ecologist whose ideas are cited as emblematic of ecofascism, a racist, xenophobic, and increasingly violent mix of environmentalism, white nationalism, and antihumanism. The idea that life is good is so commonsense in Western thinking as to rarely require justification. However, viewing life itself as the *ultimate* good can lead to dark places indeed.

A Place Person

Born to a wealthy family in Oslo in 1912, Arne Næss spent the first half of his life as a philosopher. Næss specialized in theories of language, melding set theory and hermeneutics, and he was a key figure in the Oslo school, a group of philosophers and social scientists who developed an empirical approach to semantics. Næss was interested not just in abstract approaches to language but in its practical applications as well. For example, his textbook *Communication and Argument*, which for many years was required reading for all students at the University of Oslo, provided a logical foundation for effective interpersonal communication as well as normative principles for public debate.[9] Næss retired from university life in 1969 at the age of fifty-seven, and it was then when the second act of his life really began.

Næss had long been drawn to natural places, and he was inspired by the Scandinavian tradition of *friluftsliv* (free air life).[10] Associated with outdoor recreation, such as hiking, camping, and mountaineering, *friluftsliv* also names something, well, *deeper* than that: "a way of thinking and being in nature" that is tied to the self-image of many Scandinavians as a "nature loving people."[11] From an early age, Næss spent summers at a family cabin in southern Norway in the Hardangervidda, the largest alpine plateau in Europe. A lifelong mountain climber, Næss felt a keen affinity with the mountains there, Mount Hallingskarvet in particular, where, as an adult, he built a spartan cabin he called Tvergastein, which means "across the stones." Tvergastein was not just the location where Næss would regularly retreat, disappearing into the writing of Spinoza and Gandhi and writing some of his most influential work; it was also the inspiration for that work. He wrote lovingly of the tiny flowers of the arctic landscape high above the tree line, the lichens that reminded him that even the "barren" mountaintops were "teeming with life," the mice who shared his home ("welcome" in the main room but "never invited to the kitchen"), and the occasional reindeer herd that would wander through the area, sometimes resting just outside his front door.[12]

Næss's writing about Tvergastein reveals an appreciation for humble beauty and his admiration for the persistence of life in such an extreme place. The eminent philosopher who had been knighted by the king of Norway could often be found belly down on the rocks, marveling at minute larvae on the leaves of plants less than an inch tall. Tvergastein is where Næss first understood "what it means to belong to a place," the place where he first began to think of himself as "a place-person." He came to identify so thoroughly with Tvergastein, in fact, that he considered changing his name to Arne Tvergastein, in keeping with an old Norwegian tradition of rural laborers who took ancient farm names as their own.[13]

Næss coined "deep ecology" in a speech to the Third World Future Research Conference, which was held in Bucharest in 1972. The conference was sponsored by the World Futures Studies Federation, and most of the discussion there centered on *The Limits to Growth*, a report commissioned by the Club of Rome that used computer models to predict a collision between human population and the end of natural resources. Published just a few years after Paul Ehrlich's 1968 bestselling *The Population Bomb*, *The Limits to Growth* contributed to a kind of apocalyptic neo-Malthusianism in the air in the early 1970s; population control and environmental repair were often spoken of in the same breath and by the same people. The talk in Bucharest focused on technocratic solutions to the world's problems, with little attention to social or economic issues or the ethical or political implications of population control in the developing world, which Michelle Murphy (Métis) explains is "typically narrated as the problem of too many black and brown people."[14] Næss and his colleague Johan Galtung, from the Peace Research Institute in Oslo, used their speeches in Bucharest to bring some of these issues to the table.[15] Galtung critiqued *The Limits to Growth* as "an ideology of the middle class" that ignored the needs of the world's poor. While Næss, like Galtung, spoke of the importance of social and economic issues in his speech, he focused on ecological issues and specifically the mindset that was driving the terms of global debate. The mainstream environmental movement, especially as it intersected with government and industry, tended to focus on issues like pollution and "improving human health and affluence for those in developed countries." This approach Næss described as "shallow."[16]

The "deep" ecology movement, in contrast, consisted of a much broader and deeper set of beliefs that grounded a more radical set of political commitments. Although deep ecologists, like their "shallow" counterparts, were committed to fighting pollution and resource depletion, Næss described them as motivated by a relational view of humanity and the natural world and a belief in the inherent value of diversity, symbiosis, and complexity. Deep ecologists rejected the "man in environment image" in favor of the "relational, total field image" and a "biospherical egalitarianism" that insisted on the inherent worth of all living things. These tenets of deep ecology, Næss explained, are "clearly and forcefully normative" and ought to be realized in the form of personal, ecologically minded philosophies he called "ecosophies."[17] While ecology names the scientific study of how "all things hang together" and environmental philosophy was a descriptive field in a "university milieu," an ecosophy names an everyday way of living in harmony with nature on the basis of "one's own personal code of values and a view of the world which guides one's own decisions." An individual's ecosophy would vary according to differences in the "'facts' of pollutions, resources, population, etc., but also value priorities."[18] One could thus have a Christian ecosophy, a Buddhist ecosophy, or an ecosophy based on a particular school of thought, political ideology, or the needs of a particular place. What deep ecology offered, then, were several basic principles that would provide a "unified framework" for individual ecosophies.

Næss offered sketches of the "norms and tendencies" of the deep ecological perspective in his Bucharest speech, which was published the following year as "The Shallow and the Deep, Long-Range Ecology Movement."[19] A revision and expansion of those norms was published nearly a decade later, in 1984, in the *Earth First!* journal, where they were put forward as "platform" principles of the deep ecology movement. Næss developed the principles with George Sessions, an American environmentalist, mountain climber, and professor of philosophy who was one of deep ecology's most prolific advocates until his death in 2016. Their goal was to move the platform of deep ecology from "slogans and poetry" to a "more literal formulation," they wrote, but more importantly, they endeavored to create a set of principles that would "appeal to a great many people."[20] After outlining the principles, Sessions and Næss added commentary, explained their rationale in detail, connected several principles to wider social and political issues, and explained their choice of terminology:

The Basic Principles of Deep Ecology

1. The well-being and flourishing of human and nonhuman Life on Earth have value in themselves (synonyms: inherent worth, intrinsic value, inherent value). These values are independent of the usefulness of the nonhuman world for human purposes.
2. Richness and diversity of life forms contribute to the realization of these values and are also values in themselves.
3. Humans have no right to reduce this richness and diversity except to satisfy vital needs.
4. Present human interference with the nonhuman world is excessive, and the situation is rapidly worsening.
5. The flourishing of human life and cultures is compatible with a substantial decrease of the human population. The flourishing of nonhuman life requires such a decrease.
6. Policies must therefore be changed. The changes in policies affect basic economic, technological, and ideological structures. The resulting state of affairs will be deeply different from the present.

7. The ideological change is mainly that of appreciating life quality (dwelling in situations of inherent worth) rather than adhering to an increasingly higher standard of living. There will be a profound awareness of the difference between big and great.
8. Those who subscribe to the foregoing points have an obligation directly or indirectly to participate in the attempt to implement the necessary changes.

The principles are clearly arranged to build on each other. The first principle begins by stating unequivocally the intrinsic value of "Life on Earth," rejecting an anthropocentric view of the world, establishing "human interference with the nonhuman world" as a "rapidly worsening" problem, and declaring that change is necessary. Notably, the principles' declaration of the intrinsic value of all life-forms is one of the most controversial points in deep ecology, yet it's not defended or justified in either the principles or the commentary that follows, except for a tautological note that "inherent value . . . is common in deep ecology literature."[21] The remaining principles, which continue to lay out problems and solutions, are capped with a political commitment to their implementation.

While not new by any means, what deep ecology's principles offered was a direct challenge to the anthropocentrism that held sway in mainstream, "shallow" Western environmentalism and a commitment to radical change on both a personal and a social level. Such a shift in consciousness, Sessions and Næss argued, would lead to support for broad-based, global policy changes that would create a future "deeply different from the present." Despite the talk of policy as outcome, however, only one specific policy proposal is named in the principles: population control. If human and nonhuman life is to flourish, the fourth principle explains, a "substantial" decrease in human population is necessary. Remarkably, this principle is the only one of the eight not to be accompanied by an explanation, which suggests that Sessions and Næss could count on the readers of *Earth First!* to support it.

Many of the principles echo points from Næss's original seven characteristics of deep ecology in "The Shallow and the Deep," though in a different order and with different points of emphasis. However, there is one notable omission from his original speech in Bucharest. In that speech, Næss argued that deep ecology had an "anti-class posture" attuned to the "exploitation and suppression of certain groups," speaking favorably not just of the diversity of life forms but of the diversity of human life and culture as well. Deep ecologists, Næss explained, "support the fight against economic and cultural, as much as military invasion and domination, and they are opposed to the annihilation of seals and whales as much as to that of human tribes or cultures."[22] While Næss suggested that deep ecological principles ought to extend "to any group conflicts, including those of today between developing and developed nations," he also urged "extreme caution towards any over-all plans for the future, except those consistent with wide and widening classless diversity."[23]

Not only is Næss's affirmation of local autonomy, economic justice, and critique of militarism, imperialism, and genocide absent in the revised principles, but the commentary includes a noticeable revision from the Bucharest speech, which critiqued exploitation and interventionism: "whereas 'self-determination,' local community,' and 'think globally, act locally,' will remain key terms in the ecology of human societies," Sessions and Næss write in their explanation of the principles, "nevertheless the implementation of deep change requires increasingly global action—action across borders." They then state that governments of "Third World countries (with the exception of Costa Rica and a few others) are uninterested in deep ecological issues." When the governments of "industrial societies try to promote ecological measures through Third World Governments, practically nothing is accomplished," they complain, which requires the intervention of nongovernmental organizations, which are able to sidestep governmental intrusion to "act globally 'from grassroots to grassroots.'" While not supporting military intervention, the revised principles are interventionist nonetheless and specifically name the developing world as the place for that intervention. "Strangely," writes Peder Anker, "no evidence suggests that the most original aspect of [Næss's] paper, its eco-centrism, raised any interest in Bucharest."[24] And yet, equally strangely, many of the points of social and economic critique in the Bucharest speech were omitted from the revised principles in favor of the ecocentric and biocentric worldview for which deep ecology became best known.

The terms "ecocentrism" and "biocentrism" are often used interchangeably, and many people in the deep ecology movement use one or the other to describe its orientation. Western biocentrism is associated with figures like Albert Schweitzer (for whom the reverence of life formed the basis of an elementary, universal sense of morality), Kenneth Goodpaster (whose 1978 essay "On Being Morally Considerable" claims that "being alive" is the only "plausible and nonarbitrary criterion" for moral standing), and Paul Taylor (whose book *Respect for Nature* extends moral consideration to all living things because they have a good of their own).[25] An ecocentric ethic, in contrast, places whole ecosystems at the center of consideration, which includes biotic as well as abiotic elements of the natural world: rocks, snails, streams, soil, microorganisms, storms. In deep ecology, the frequent conflation of these terms might have its origin in Sessions and Næss's initial commentary on the platform principles. There they explain that their use of the term "life" denotes a broad, "more comprehensive non-technical way to refer also to what biologists classify as 'non-living' rivers (watersheds), landscapes, ecosystems." "Life," they continue, thus refers to "the biosphere, or more accurately to the ecosophere as a whole. This includes individuals, species, populations, habitat, as well as human and non-human cultures."

Yet despite Sessions and Næss's ecocentric focus, life and living things are given repeated positions of emphasis in the principles and the commentary. Note how "Life on Earth" is capitalized in the principles, how only human "vital" needs are sufficient to intervene in nonhuman nature. In the commentary on the second principle, for example, Næss and Sessions explain that it "presupposes that life itself, as a process over evolutionary time, implies an increase of diversity and richness." While "evolutionary" might be used as a synonym for deep time and thus speak to the ever-changing forms of rivers and mountains, the simplest explanation is that "life itself" refers to the totality of biological life-forms subject to natural selection. If the concern of deep ecology, then, is ultimately and "accurately" with the "ecosphere as a whole," then why don't Sessions and Næss say *that*, rather than "Life on Earth," which they then need to qualify?

When viewed as a rhetoric, what first appears to be inconsistency in the principles takes shape as a strategy: deep ecology promotes an ecocentric worldview using biocentric *rhetoric*. Despite Sessions and Næss's goals to use the principles as a "literal formulation" of deep ecology, the meaning of "life" in the principles is deliberately "non-technical," they write, reflecting a "broader usage" that would encompass "slogans such as 'let the river live.'" While in chapter 1, we saw how the meaning of "life" in some Indigenous cosmologies extends to entities such as waterfalls or stones, the reason Sessions and Næss use "life" here is not a statement of ontology so much as it is a description of deep ecology's rhetorical approach. So what, then, does "life" *do* for deep ecology? One answer may be found in one of its central concepts: identification.

Identification, Self, and Other

Identification and self-realization are, "without question, the two concepts most often associated with the views of deep ecology supporters," explains the environmental philosopher Christian Diehm.[26] One of the fundamental goals of deep ecology is to broaden one's identification with the other-than-human world, which in turn results in a more expansive sense of self: from the bounded, individual self to the wider, "ecological self" that extends beyond the individual human body to include birds and grizzly bears, grasses and bacteria, mountain ranges, weather, and watersheds.[27] The ecological self, Næss explains, is simply "that with which this person identifies. This key sentence (rather than definition) about the self, shifts the burden of clarification from the term 'self' to that of 'identification' or 'process of identification.'"[28] In deep ecology, identification means "what we ordinarily understand by that term," writes the philosopher Warwick Fox, "the experience not simply of a sense of similarity with an entity but of *a sense of commonality*."[29]

One of the most detailed discussions of identification is found in Fox's book *Toward a Transpersonal Ecology*, an "exhaustive" account of, and argument for, deep ecology as an ecophilosophy.[30] Building on Næss's notion of the ecological self, Fox explains that deep ecology encourages its adherents to shift from personal to "transpersonal" identification, a term he takes from psychology. Personal identification names "experiences of commonality with other entities that are brought about through personal involvement with these entities."[31] The entities with which we might personally identify range from the concrete, such as members of our family, friends, pets, and even toys, to the abstract, such as clubs or nations or other communities to which we belong. Personal identification, Fox suggests, leads people to see their interests as inextricable with the interests of those with whom they identify. An attack on them is an attack on us; to defend them is to defend ourselves. Personal identification proceeds from the inside out, Fox explains, from the self to others: from an individual person to that which is physically or psychologically closest to that person and only then to a broader "sense of commonality with other entities."[32]

Transpersonal identification, in contrast, works from the outside in. Fox names two types of transpersonal identification: *ontological* identification (a focus on nondual Being found in philosophers like Heidegger and in some schools of Zen Buddhism) and *cosmological* identification, which defines Næss's thinking in particular. Cosmological identification refers to the "experiences of commonality with all that is that are brought about through deep-seated realization of the fact that we and all other entities are aspects of a single unfolding reality."[33] Fox notes that the cosmological identification might be found in the thinking of certain Indigenous peoples, religious traditions like Taoism or Hinduism, some schools of philosophy, and some domains of science, particularly when they direct our attention to the interconnections between humans and the other-than-human world.

Diehm argues that transpersonal identification might be split even further into two forms, each of which emphasizes a slightly different meaning of the term. The first, "identification-as-belonging," challenges the individualism central to Western visions of humanity in favor of a relational model that positions humans as "*belonging to* or [in] *community with* the other-than-human world."[34] This is the "relational, total field image" of the human that Næss described in his Bucharest speech but is also akin to what Aldo Leopold describes in his land ethic, which calls for a shift in the self-concept of *Homo sapiens* from a "conqueror of the land-community" to a "plain member and citizen of it."[35]

The second type of identification that Diehm finds in deep ecology he names "identification-as-kinship." Although this type of identification is less commonly discussed in deep ecology discourse than identification-as-belonging is, he argues that "it is vital to Næss's ecosophy, and arguably is the principal sense of the term as he understands it."[36] While identification-as-belonging involves experiencing a sense of *self* in an expanded, ecological, relational way, identification-as-kinship involves experiencing *others* or "recognizing something of importance about *them*. It is only after, so to speak, that there are consequences for the self; and that there are such consequences is because this form of identification allows us to relate to others *on such terms as to be able to care for them deeply*—so deeply, in fact, that our very self is implicated."[37]

What is at stake in identification-as-kinship is the question of what motivates solidarity and care for the other-than-human world: Is it similarity or difference? This is a point that several critics of deep ecology have seized on, charging that in its focus on self-realization and the ecological self, deep ecology fosters ecological thought and political action based on a self- rather than an other-oriented approach to ethics. In an incisive 1989 essay, for example, Peter Reed argues that the existential divide between (Western) humanity and the other-than-human world should be embraced, not overcome: "it is our very *separateness* from the Earth," he writes, "the gulf between the human and the natural, that makes us want to do right by the Earth."[38] Drawing on Martin Buber's I-Thou relationship and Rudolf Otto's idea of the Holy, Reed argues that the other-than-human world should be considered not as part of the self but as a "stranger," the "Wholly Other," "a self-sufficient being" of whom we have but "an inkling." In this way of thinking, the other-than-human world is not just outside Western ethical categories such as intrinsic value or rights but beyond, indeed *above*, them. The natural world is superior to the human world, Reed argues, the grounds for a sublime relationship based in awe and reverence that also "inevitably arouses a feeling of obligation." Ultimately, he argues that "we can only experience the mysterious otherness of nature through meetings with a dominant nature" that must be beheld and respected.[39]

Extending Reed's critique, the ecofeminist Val Plumwood also argues for the value of difference in motivating solidarity with the other-than-human world. However, rather than flip from a self-based to a *wholly* other-based approach, she argues for a more nuanced position that builds on insights from feminist and postcolonial theory. Noting that "the drive to hyperseparation is part of the colonizing conceptual dynamic that places the human colonizer radically apart from, and above," a subordinated nature, Plumwood argues for a "concept of the other as interconnected with self, but also as a separate being in their own right," maintaining tension between similarity and difference while also preserving a sense of distinction, which is a point I will return to in chapter 3. In deep ecology's focus on the self, it was poised to join "oppressive projects of unity," Plumwood argues.[40] These projects are especially apparent in relations of settler colonialism, such as settlers' attempt to lay claim to Indigenous identity or culture or the effort to assimilate Indigenous people into settler culture through acts of cultural genocide like residential schooling.[41] Oppressive projects of unity—which grow out of the "incorporative self of the colonizing mind"—position the self *as* the other, whereas solidarity positions the self *with* or in support of the other.[42] This aligns with the difference between what Gayatri Spivak describes as the call *to* the ethical and the call *of* the ethical. The latter, writes Karma Chávez, means that one responds to the other on their terms, rather than one's own, and implies "a sense of responsibility to an other who is different."[43]

Diehm argues that identification-as-kinship serves as a counter to critics who suggest that in expanding the self, deep ecology is merely an expansion of self-concern or that it erases difference and exfoliates distinctiveness in its pursuit of unity. To illustrate this idea, he turns to a story that Næss frequently used as a "paradigm situation" of identification: an encounter with a "nonhuman being" he met in the 1940s. Næss recalls,

> I was looking through an old-fashioned microscope at the dramatic meeting of two drops of different chemicals. At that moment, a flea jumped from a lemming that was strolling along the table. The insect landed in the middle

> of the acid chemicals. To save it was impossible. It took minutes for the flea to die. The tiny being's movements were dreadfully expressive. Naturally, I felt a painful sense of compassion and empathy. But the empathy was not basic. Rather, it was a process of identification: I saw myself in the flea. If I had been alienated from the flea, not seeing intuitively anything even resembling myself, the death struggle would have left me feeling indifferent. So there must be identification for there to be compassion and, among humans, solidarity.[44]

For Diehm, Næss's story of the flea is not necessarily an expression of belonging or interconnectedness. Rather, it "involves catching sight of commonality or continuity between ourselves and other entities."[45] For Næss, this commonality could take the form of capacities or vulnerabilities that we share, common needs or interests, the capacity for suffering, or even "that part of God that lives in all that is living," an echo of Ralph Waldo Emerson's sublime experience in *Nature*.[46] For Næss, identification *precedes* empathy or sympathy, Diehm explains, and emerges from an a priori connection, as opposed to alienation. In this respect, it echoes a point that Diane Davis makes about Burkean identification: that despite its emphasis on division, "identification could not operate among self-enclosed organisms; it would have to belong to the realm of affectable beings, infinitely open to the other's affection, inspiration, alteration."[47] In this sense, we might read identification in Næss as affectability itself—an openness to the other that makes connection possible.

Diehm's analysis is a helpful guide to the forms that identification takes in deep ecology considered as a philosophy. When viewed as a rhetoric, it also invites a number of questions. For example, the meaning of kinship is assumed in Diehm's description rather than defined or argued for. Using the article's context as a guide, I gather that it comprises two aspects in deep ecology: an other-oriented perspective and a sense of commonality, which motivate both obligation and care. What is left unaddressed is how that happens and to what end. As Kyle Whyte explained in our conversation in the previous interchapter, for example, kinship also might be seen as a moral bond that extends beyond those who are immediately related by it—a relation of mutual responsibility grounded in community purpose.

Donna Haraway offers a slightly different meaning of the term "kinship" in her book *Staying with the Trouble*. Kin is "an assembling sort of word," writes

Haraway; kinship is a process that "stretches" and "recomposes" relations, and it arises from the "fact that all earthlings are kin in the deepest sense." Humans, she argues, need to learn not only how to "recognize" but also how to "make" human and other-than-human kin—a political urgency in the current moment, when bonds of all kinds seem to be unraveling. But if not with genes or blood, then with what is kinship's connective tissue made? "What shape is this kinship, where and whom do its lines connect and disconnect, and so what?" And to what end? Haraway suggests a diffuse "multispecies flourishing" as one outcome but also argues that making kin is the process of "making persons, not necessarily as individuals or as humans."[48] Haraway's use of "person" here summons the subject of Western ethical and legal discourse; persons are those to whom moral consideration and legal protection may be granted, accompanied by the knot of rights and responsibilities that this status entails.

Broadening personhood beyond humans is a form of ethical extensionism, the effort to expand moral standing beyond the human, and it is a familiar approach in both environmental and animal ethics. Finding the restriction of moral considerability in most ethical theory to be arbitrarily based on the human species, some extensionist approaches instead locate moral standing in qualities and capacities that have expanded in wider and wider circles over the course of Western history—from the self to the family to the group to the nation to all humans and, as we are seeing in recent years, to some animals.[49] This move of shifting value across entities, or what Allison Rowland describes as a "zoerhetoric," elevates some nonhuman animals like great apes and dogs to moral patients, but in the process, it also troublingly demotes some human beings, such those with certain disabilities, below the threshold for personhood.[50] Diehm explicitly notes that identification-as-kinship could be regarded as a type of extensionism, and this might be the grounds of a "serious objection" to it, given that many forms of extensionism suffer from an anthropocentric bias, using criteria such as sentience or sapience for moral standing.[51] Yet sentience or sapience are not the only qualities that have or can be used as the grounds for moral considerability—life, for example, is also used as a criterion for moral standing in a biocentric approach to ethics. But if we look even closer at the identification of deep ecology, we see that life is employed as more than just the grounds for moral standing or as a substance held in common between the human world and the other-than-human world. By invoking a sense of collective precariousness, bioidentification arouses a feeling of obligation and inspires practices of care.

Precarious Life

Let's return to the example that Diehm uses to illustrate Næss's approach to identification-as-kinship: his encounter with the flea in the laboratory. Insects are the tiniest animals with which we are intimately familiar and whose lives we are well acquainted with taking—often en masse. They are the quintessential form of life whose life is considered so insignificant that it may be taken without a second thought or even a first one. In many cultures, there are laws in place to prevent a litter of kittens from being drowned or to decrease the needless suffering of cattle during slaughter. But the destruction of insect life rarely merits reflection, much less grief or accountability. You may have judged me for killing that fruit fly *but only a little bit.*

From a rhetorical perspective, it *matters* that Næss uses a flea to illustrate how identification works and what it entails, as opposed to, say, an orangutan.[52] Næss sees the flea's struggle for life, and it arouses a "sense of compassion" that he links to "a process of identification: I saw myself in the flea. If I had been alienated from the flea, not seeing intuitively anything even resembling myself, the death struggle would have left me feeling indifferent."[53] To be sure, one could fairly apply Plumwood's and Reed's critique to Næss's flea encounter; it's not difficult to see why a self-oriented approach to ethics—a throwback to what Roderick Nash describes as a "pre-ethical" past—is troubling.[54] If one is moved to act only when one's own interests are threatened or only on behalf of similar others, one risks remaining "indifferent," as Næss puts it, to the suffering of those who are perceived to be different or whose interests do not intersect with our own. This is one way that atrocities happen and also how atrocities are justified by those who, despite all evidence to the contrary, still consider themselves to be good people.[55] But I think something more is going on here.

To illustrate identification in *Ecology, Community, and Lifestyle,* Næss once again provides a narrative example using insects:

> In a glass veranda with one wall open away from the sun a bunch of children are playing with an insect spray. Insects are trapped flying against the wall pointing towards the sun. Spraying makes them dramatically fall to the floor. A grown-up appears, picks up an insect, looks at it with care, and wonders dreamingly: "perhaps those animals might, like you, prefer to live rather than to die?" The point is grasped, the children for a moment to see and experience spontaneously and immediately the insects as themselves,

> *not only as something different but in an important sense like themselves.* An instance of momentary identification! Perhaps it has no effect in the long run, or perhaps one of the children slightly changes in attitude toward small fellow creatures.[56]

Presented as a hypothetical, this example is composed with a clear purpose: to illustrate Næss's approach to identification and how it works. The adult doesn't tell the children to stop their insecticide, nor does he offer a deontological principle with which they can deduce that what they're doing is wrong. Instead, the adult seeks to evoke *fellow feeling* in the children on the basis of a shared preference for life and a shared aversion to death—a commonality that Næss believes might compel a change in attitude "toward small fellow creatures" in the future, however slight. The adult does not tell the children that they are the *same* as the insects ("*not only as something different* but *in an important sense* like themselves") but seeks to evoke a sense of commonality.

If you recall, earlier Diehm noted that the identification-as-kinship that characterizes Næss's deep ecology means to "realize that others are enough *like* we are to command the same sort of concern and respect that we have for ourselves." And yet I think this is not *quite* what Næss is doing here, even though he uses similar language in both examples. The life that Næss presents or, more appropriately, the *striving* for life and against death that he identifies (what Spinoza describes as *conatus,* the struggle of a being "to persist in its own being")[57] in his insect examples is not used as an analogy. It is not something that is characteristic of an individual or the human as such that is then extended to the insect (in the way that, for example, sentience or sapience might be). Rather, what Næss attempts to evoke in these examples is a sense of life in the other as precarious, to use Judith Butler's term, which then folds *back* on the self, as Diehm suggests.

While Butler is primarily concerned with the precarity of human subjects in her books *Precarious Life* and *Frames of War,* Joshua Trey Barnett argues that this work provides a "generative" framework for scholars interested in the complexities of "earthly co-existence" with the other-than-human world.[58] Butler draws heavily from Levinas in her writing on precariousness, especially his notion of the face. To fully delve into the nuances of Levinas is beyond the scope of this chapter, and I'm not interested in wading into the controversy over whether his notion of the face applies beyond the human, especially when others have made that case thoroughly elsewhere.[59] But questions about animal faces aside, one point that Butler takes from Levinas seems worth considering in its relation to Næss's bioidentification. Butler explains that to "respond to the face, to understand its meaning, means to be awake to what is precarious in another life, or, rather, *the precariousness of life itself*. This cannot be an awakeness, to use his word, to my own life, and then an extrapolation from an understanding of my own precariousness to an understanding of another's precarious life. It has to be an understanding of the precariousness of the Other."[60]

The zoerhetorical alchemy that makes an entity "intelligible *as a life*," in Butler's terms, happens "spontaneously," as Næss puts it.[61] Bioidentification works in deep ecology not via an orderly set of premises that proceed from the self to the other (I want to live; I am worthy of moral consideration; this being wants to live; this being is worthy of moral consideration) or deontological deduction (thou shalt not kill) but as an apprehension of life itself as shared and precarious, an immediate, affective cascade that draws the self and the other into ethical relation. As Butler explains, "precariousness implies living socially, that is, the fact that one's life is always in some sense in the hands of the other."[62] As Barnett explains, however, the social interdependency that Butler identifies in precariousness can be extended to the ecological as well. When Butler describes our lives as in the hands of the other, he writes, "It is not just 'hands,' and certainly not just human hands, that we depend upon, but rather, an entire teeming ecosystem."[63] As such, when she writes that "precisely because a living being may die, it is necessary to care for that being so that it may live," we might extend this sense of fragility to the rest of the living world.[64] Just as we are exposed to the world by the fact of being alive, so too is it exposed to us, as Barnett points out: "Precariousness entails a certain fragility . . . as a constitutive component of what it means to be a living body exposed to the world in which one finds oneself."[65] "That which is fragile," argues Robert Cox, "requires protection or an agent's active intervention to ensure its continued existence."[66] But as Barnett reminds us, sometimes we will also "have to learn to *stop* intervening precisely because we care."[67]

While most evident in Næss's writing, bioidentification is found elsewhere in deep ecology too. Consider, for example, Fox's discussion of identification in *Toward a Transpersonal Ecology*. If you recall, Fox's notion of transpersonal identification includes both ontological and cosmological identification, which he associates with Næss and which Fox describes as "feel[ing] a sense of commonality with all other entities."[68] While Fox explains that cosmological identification is not "theoretically" preferable to ontological identification—both are modes of transpersonal ecology—he explains that the former is "practically" preferable because it "can readily be inspired through symbolic communication."[69]

To illustrate how identification works, the telling metaphor that Fox reaches for is the tree of life.[70] Fox prefers this image for several reasons: the "leaf" image suggests similarity as well as difference, and its common use as a genealogical metaphor evokes a sense of relation that extends into the deepest reaches of deep history: it "captures the fact that all that exists has arisen from a single seed that has grown into an infinitely larger and infinitely more differentiated entity over time." The tree of life metaphor explains not just how existing entities are related in the present, in other words, but how those relations might be traced back to the initial "seed of energy" that, fourteen billion years ago, gave rise to all that is. But what is most important to Fox is how the tree of life metaphor communicates the relationship between vulnerability and care:

> The image of leaves on a tree clearly suggests the existence of an entity that must be nurtured in all its aspects if all its aspects are to flourish. Damage a leaf badly enough and it will die; damage a branch (say, the branch of cosmic differentiation that became the Earth some 4.6 billion years ago) and all the leaves on that branch will die. In contrast, the images of drops in the ocean, modes of a single substance, ripples on a tremendous ocean of energy, and knots in a cosmological net do not readily suggest the existence of entities that need to be nurtured in any way at all.[71]

Fox explicitly prefers the tree of life metaphor over others not because it is more accurate but because of the ease with which it communicates precariousness and the need for care. Næss could have used stones, but he didn't. Fox could have used drops in the ocean, but he didn't. Both use identification with living things to create a sense of commonality in the precariousness of life itself, as well as the sense of obligation and practice of care that serves as the foundation for political action.[72] While the sense of commonality that is the heart of deep ecology has been critiqued as self-centered at best and self-indulgent at worst, it is clear that it serves a clear rhetorical and political purpose. Fox concludes his book by noting that identification is expressed in personal actions that "tend to promote symbiosis," but also in "actions that respectfully but resolutely attempt to alter the views and behavior of those who persist in the delusion that self-realization lies in the direction of dominating the earth and the myriad entities with which we coexist."[73] Deep ecology's metanoia, in other words, does not involve simply changing one's own mind; it is an active call to change the minds of others on behalf of, and as a part of, the fragile, living world.

Apprehending precariousness means to see one's own hands as the conditions by which the other's life may flourish or be harmed, but as Barnett argues, it is also to appreciate the conditions of one's own flourishing or harm in the hands (and the paws, leaves, cilia) of countless others.[74] However, while identification-as-kinship evokes a lovely image of harmonious interconnection, it is crucial not to overlook deep ecology's point in the platform principles that the hands that have wrought ecological destruction are *human* hands. This point has led to distinct strains of antihumanism in some quarters of deep ecology, the subject of several piercing critiques by the great green anarchist Murray Bookchin in the 1980s. In his essay "Social Ecology versus Deep Ecology," for example, Bookchin charged that many radical environmentalists were "barely disguised racists, survivalists, macho Daniel Boones, and outright social reactionaries" whose antihumanism portended a brutal politics.[75] Peter Reed, for example, who critiqued deep ecology's centering of the human self, denounced his fellow humans as "terrorists" who "hold the world hostage." Faced with the "clash" between human interests and the "survival of the planet's nature," he wrote, "I think it is humans who must give way." To explain what he meant by this, Reed turned to an infamous metaphor from the ecologist Garrett Hardin. At a time of limited resources, growing human population, and inevitable "ecoholocaust," Reed wrote, characterizing Hardin's position, "if poorer or stupider nations drown under a tide of their own humanity, that is their own fault. Rich 'lifeboat' nations should not jeopardize their own survival by helping the suffering. Compassion for the victims means catastrophe for the race." Reed admitted that this position sounds "inhuman," but he claimed that it was ultimately justified by Hardin as necessary to ensure a future for the human species. Reed described his own proposal as "even more ghastly than Hardin's." For the sake of "the austere mystery of nature as *Thou*," he wrote, "we have to ask ourselves whether the preservation of humanity is of overriding importance."[76] Bookchin's concern with the antihumanism of environmental advocates like Reed was not just that it was "counter-productive" to the movement's goals but that it posed a distinct danger. "I saw this happen in the 1930s," he argued, pointedly commenting that all of his European relatives had been murdered in the Holocaust. "That is why I say that eco-fascism is a real possibility within our movement today."[77]

Taking Lives to Save Life

Humankind alone is no longer the focus of thought, but rather life as a whole. . . . This striving toward connectedness with the totality of life, with nature itself, a nature into which we are born, this is the deepest meaning and the true essence of National Socialist thought.
—Ernst Lehmann, *Biologischer Wille*

Environmentalism is often considered the domain of the political left, but this is not necessarily the case. Some conservative politicians, for example, have championed conservation efforts, and although it seems hard to believe after decades of Republican climate denialism and environmental deregulation in the United States, the establishment of Earth Day was initially a bipartisan initiative. Likewise, several positions now associated with the far right may be found in the rhetoric of radical environmentalists ostensibly on the left. Dave Foreman, for example, was eventually pushed out of Earth First! for his increasingly nativist views on immigration.[78] And Edward Abbey, author of environmental classics like *Desert Solitaire* and *The Monkey Wrench Gang*, once penned a racist essay that condemned Mexican migration, in which Abbey claimed, using the lifeboat trope once again, that the "United States has been fully settled, and more than full, for at least a century. We have nothing to gain, and everything to lose, by allowing the old boat to be swamped" by "millions of hungry, ignorant, unskilled, and culturally-morally-genetically impoverished people."[79]

Recent years have seen an even more ominous link between white supremacy, nativism, and environmentalism in the emergence of the ecofascism that Bookchin feared, whose adherents have advocated—and used—violence as a means of advancing their worldview. Though a full history of this nascent movement is outside this chapter's scope, its genesis is often located in Nazi Germany, which promoted a mythical, essentialist view of the German "race" tied to nature and territory found in the phrase *Blut und Boden*—that is, "blood and soil," the same phrase chanted by torch-wielding white nationalists at the deadly "Unite the Right" rally in Charlottesville, Virginia, in 2018.[80] In the first two decades of the twenty-first century, ecofascism has emerged as a counterpart to growing movements of ethnonationalism in the United States and Europe. These ideologies have flourished in part because of digital forums like 8chan as well as print forums like the Budapest-based publisher Arktos Media, which features books with titles like *The Blackening of Europe, A Handbook of Traditional Living, Understanding Islam,* and *A Handbook for Right-Wing Youth.*

In 2011, Arktos published a book by a Finnish ecologist named Pentti Linkola titled *Can Life Prevail? A Revolutionary Approach to the Environmental Crisis.* Calling Linkola "among the foremost exponents of the philosophy of deep ecology," the publisher describes the book as "a refreshing departure from the usual rhetoric on this topic."[81] While it might be tempting to write off Linkola as a fringe thinker whose literature is passed hand to hand at neo-Nazi rallies, I will note that this book is for sale on Amazon (free to read on Kindle Unlimited), and as I write this, it has nearly seventy positive ratings—an average of five stars. Reviews praise Linkola as an "original" and "creative" thinker, "ruthlessly honest," and recommend the book as one that will "challenge your existing mindset, provoke a reaction, and make you think."[82]

A casual reader of *Can Life Prevail?* might find what initially passes as a collection of environmental essays written by a cranky Finnish fisherman afflicted with nostalgia for the premodern world. (It's no surprise that Amazon suggests that readers of Linkola have also read works by Theodore Kaczynski, otherwise known as the Unabomber, who has attracted a growing fan base among young people in recent years.) The book features arguments in favor of sustainable farming, rants against highways and foresters, laments for the lack of a viable green party in Finland, concern for the health of bird populations, praise of "virgin forests," and two separate essays lambasting the domesticated cat.

It is in Linkola's commentary about cats, in fact, where the grimmer side of his worldview first comes into focus. Linkola emphasizes several times that the cat is not native to Finland, calling it "an angel of death imported from Egypt." Linkola's hatred of cats is tied to what he sees as their uncontrollable fecundity: "I am not sure how many years it would take cats to cover the face of the Earth," he suspects, "but it wouldn't be many."[83] Other nonhuman animals are also given the same "foreign" treatment, such as minks (Canada) and raccoons (China).[84] Linkola explains that it might be possible to "tolerate the importation of alien species as long as they do not harm the native ones. But if the existence of any native species is threatened in order to secure the well-being of imported animals—if goshawks are threatened because of pheasants, for instance, or lynxes because of white-tailed deer—then the environmentalist's verdict must be irrevocable."[85] Because of their threat to "original, natural animals," Linkola explains, "these animals (the mink, raccoon and cat in Finland) will have to be stripped of all rights."[86]

Linkola's concern with invasive species and even his villainization of cats are common issues in environmental discourse and policy.[87] However, when set next to the other themes of *Can Life Prevail?*, a much more ominous picture takes shape. There is a long history of rhetorical cross-pollination that draws together discourses on invasive species, immigration, and nationalism.[88] As Banu Subramaniam argues, this discourse draws on a number of xenophobic parallels between human immigrants and nonhuman invasive species, positioning both as alien, silently growing "in strength and number," uncontrollably fecund, and driven by an impulse to take over.[89] While Linkola draws on many of these tropes in his discussion of nonhuman species, one doesn't need to interpret Linkola's feelings about cats to find nativism and racism under the surface. He explicitly decries human migration in similar terms, writing that "immigrants from poor nations" have a birth rate "at par with that of their cultures of origin (if not higher, thanks to the social care they now benefit from).... As Matti Kuusi once put it, there is no use counting the immigrants at the border: one should wait a while and look in their nurseries."[90] Once again, this book is published by a *white supremacist press*, whose founder is a Swedish neo-Nazi skinhead turned far-right media mogul who has partnered with the American white nationalist Richard Spencer.[91] But where Linkola departs from even the racist far right is in the problem he describes and the solution he proposes.

While Linkola notes his "love" for Finland (which he calls his "motherland"), he does not have kind words for Western culture in general, which he argues has brought both humanity and the natural world to a state of ruin.[92] While his description of nonwhite immigrants is draped in nativist, racist language, he notes that his *primary* concern is that they "dramatically increase the wealthy population and environmental burden of industrial countries." This is no mere critique of human-induced environmental degradation. Linkola paints the situation in apocalyptic terms: "the end of history is nigh," he writes, and we are on "a rapidly accelerating pace toward final suffocation."[93] For the sake of the survival of life itself, then, Linkola argues that decisions need to be made. Once again, Linkola's hated cats offer a preview. He not only argues that cats ought to be deprived of rights (in favor of the native species they threaten) but goes on to complain about a "new insane resolution" that prohibits drowning them. For "ages," humans have drowned unwanted cats, Linkola argues. "If anything, this is a humane act, considering that even in the case of humans drowning is the easiest and most blissful way to die."[94]

In the most disturbing essay in the collection, "A Refresher Course in the State of the World," Linkola provides an account of what he calls our current "ecocatastrophe" and the entity responsible: humanity. "Everywhere," he writes, "man remains a complete lout, a destroyer of the biosphere." The principal cause of the "impending collapse of the world—the cause sufficient in and by itself—is the enormous growth of the human population: the human flood. *The worst enemy of life is too much life*: the excess of human life." No human activity has meaning if there is no life on the planet, he explains. As such, he offers what he calls "the doctrine of the protection of life," which he notes is "the highest objective, all other goals being subordinate to it. . . . *The protection of life is justified at whatever cost*." For the protector of life, Linkola explains, "it is unthinkable that the whole Earth should belong only to one animal species, humanity. . . . Does man have the right to rule the destiny of millions of basically similar species? Is man the living image of God?" The guardian of life is not just motivated by his version of reason and logic. "The basic principle of life protection, the conservation of the Earth's life as a lush and diverse whole, is also perceived as being sacred: as something incomparably holier than anything man might regard as such," he explains, in an echo of Peter Reed.[95]

From these premises—that life itself is a sacred, ultimate term to which all else is subordinated, that this life is under threat by humanity, which is ever increasing its numbers—Linkola takes the reader to his inhuman conclusion, using Hardin's lifeboat metaphor, though with his own gruesome spin: "What to do when a ship carrying a hundred passengers has suddenly capsized, and only one lifeboat is available for ten people in the water? When the lifeboat is full, those who hate life will try to pull more people onto it, thus drowning everyone. *Those who love and respect life will instead grab an axe and sever the hands clinging to the gunwales*." The "human flood" must be stopped by any means necessary, Linkola states, matter-of-factly, and it is important to remember that earlier in the book he had praised drowning as a "blissful" and "easy" way to die. Aware that this position will raise alarm in his readers, Linkola waves off any potential concerns. After all, he contends, even the "massive depopulation operations of Stalin and Hitler" did not overturn our "ethical norms." Yet, he argues, "This fear endures, regardless of how elegantly the reduction of the population might take place, were it even to occur more artlessly and discreetly than with the German gas chambers during World War II—possibly by limited nuclear strikes or through bacteriological and chemical attacks against the great inhabited centres of the globe (attacks carried out either by some trans-national body like the UN or by some small group equipped with sophisticated technology and bearing responsibility for the whole world)."[96] That Linkola implicitly speaks of the Nazi gas chambers as *artful*, that he suggests the use of nuclear strikes or chemical warfare as a means of reducing the population *elegantly*—these are not lines that he includes for shock value, even if shock may be the effect. In many ways, Linkola sets up the reader for this argument. He builds to this point slowly using amplification—first laying the groundwork with anodyne praise of unspoiled forests, the menacing of native birds by foreign cats, all building to a voracious "human flood" that threatens life itself. Having crucially established life itself as the ultimate term, Linkola leads the reader to his ghastly conclusion.

"Life," Linkola writes, "is hierarchic by nature." The guardian of life, taking life in the name of life, is thus not just defending nature but doing its bidding. Citing Nazi atrocities, the main problem with "our age" is not that it does not respect human life but that it *overvalues* human life, Linkola argues. The very notion of human rights, he suggests, is a "death sentence" for the rest of the living world. While Linkola may correctly be described as antihuman, it's also clear that the hierarchy within his understanding of "life" extends to a hierarchy in humanity as well: "I could never find two people who are perfectly equal," Linkola attests: "One will always be more valuable than the other. And many people, as a matter of fact, simply have no value."[97] One is reminded here, explicitly, of the Nazi phrase *Lebensunwertes Leben*—life unworthy of living—as well as its use as a chilling rationale: those lives deemed unworthy of life were seen as a drain on the nation's resources and targeted for elimination.[98]

Ecofascism like the kind Linkola advocates in *Can Life Prevail?* has tragically not remained on the page. Two of the most high-profile mass shootings in recent memory—an attack targeting Muslims at the Al Noor Mosque in Christchurch, New Zealand, which killed fifty-one people, and an attack targeting Latinos at an El Paso Walmart, which killed twenty-three people—were carried out by young white men who explicitly claimed ecofascist beliefs. The manifesto of the El Paso murderer, for example, is titled "The Inconvenient Truth," which bemoans the pollution of watersheds, destruction of forests, plastic, and electronic waste, blaming an American culture of consumerism. Given that politicians, beholden to corporations, were reluctant to address these issues, the "next logical step," the manifesto attests, "is to decrease the number of people in America using resources. If we can get rid of enough people, then our way of life can become more sustainable."[99]

Conclusion

Arne Næss saw deep ecology as "squarely an antifascist position," and I have no doubt that he would have been horrified by Linkola's worldview as well as his alignment with deep ecology.[100] While there are similarities in their biocentric rhetoric, there are key differences between deep ecology and ecofascism, and they are found in the precariousness that Butler tells us is inextricable from all thinking about life. "In our collective precariousness," Barnett writes, extending this position, "our lives are both made possible and threatened by the others with whom we cannot help but relate."[101] For deep ecologists, this relation, and the vulnerability and obligation it summons, is the grounds for a radical politics that crosses species. But the ecofascist views life not as *collectively* precarious, not as inescapably and mutually bound together; rather, the ecofascist views life as an exclusive, zero-sum game.[102] In this relationship, it's not only that the life of a cat or a migrant does not constitute *a* life, a life that matters, but that the very existence of the other's life threatens the life of the self. And *this* is what a wholly self-oriented ethics looks like, a bioidentification that shares a meaningful life across species but limits a sense of commonality to those in perceived proximity—Finnish goshawks but not Egyptian cats, Scandinavian humans but not Syrian ones. In this nightmarish world in which life is a relationship not of mutual vulnerability and responsibility but of mutual exclusion, only the self is precarious. Ecofascists recognize that their lives are indeed "in the hands of the other," but instead of reaching out to grasp those hands, they grab an axe.

3

Death Itself | The Politics of Human Extinction

What are the limits of invoking a supposed "human species," whose relationship to itself it may rediscover only because it is exposed to the peril of its own extinction?
—Achille Mbembe, *Necropolitics*

Five times in Earth's history, cataclysmic natural disasters extinguished more than 75 percent of life on the planet. One event, the Cretaceous-Paleogene extinction, which famously killed the dinosaurs, is thought to have been caused by an asteroid. The other four were caused by a variety of factors, such as the movement of continents, volcanic eruptions, toxic algae blooms, ocean acidification, rising and falling sea levels, and, in nearly all cases, climate change. Many scientists believe we are now living during the Earth's sixth mass extinction event, the scope of which is still not fully understood.[1] Because the phenomenon is difficult to measure, estimates of the current rate of extinction vary considerably. Rates are derived from computer models, not direct observation, and documented species losses are difficult to come by. Some studies suggest that the current rate may be up to one hundred times higher than the natural background rate, the rate at which species have gone extinct throughout geological time.[2] Others put the current rate at more than a thousand times higher than the natural background rate.[3] Using a different approach, the conservation biologist Gerardo Ceballos argues that the attention to extinction (that is, total species loss) risks overlooking the full scale of the issue.[4] Focusing on the decrease of animal individuals in a given population—the precursor to extinction—Ceballos and his colleagues estimate that as many as half of the "number of animal individuals that once shared Earth with us are already gone." The loss of life on the planet, they maintain, should be seen as nothing less than "biological annihilation."[5]

While precision may be elusive, there is no doubt that we are in the midst of a crisis. Because the current extinction event is a human creation, not just a natural, cyclical process, public advocacy and political action are of utmost importance.[6] Despite the staggering implications of mass extinction, however, it remains "low on the political agenda."[7] One reason may be that extinction simply does not make the news, with the exception of the loss of a species that humans have determined is valuable or significant for some reason.[8] Though the two are related, climate change receives eight times more press coverage than biodiversity loss.[9] For policy makers, even those sympathetic to environmental issues, the extinction crisis is a classic example of a wicked problem: an issue so large and so complex that it is difficult to describe or understand, let alone act on.[10] Despite these formidable challenges, it is clear that without radical changes to the systems of energy, land use, transportation, economics, and agriculture in the wealthy West and Global North, the planet will continue to experience a loss of life of such magnitude that its full impact is difficult to comprehend. "All signs point to ever more powerful assaults on biodiversity in the next two decades," concludes Ceballos and his colleagues, "painting a dismal picture of the future of life, including human life."[11]

Though rare in scientific discourse, which tends to mute emotion and measure tone, the apocalyptic picture that the Ceballos paper paints is common in public environmental discourse. From *Silent Spring* to *The Population Bomb* to *An Inconvenient Truth* to *The Lorax*, environmental prophets have warned that short-sighted, selfish human actions will lead to a radical transformation of reality, usually portrayed as an irremediable ending—the end of life as we know it, the end of civilization, the end of the world, the end of human life, the end of life itself.[12] As Jennifer Clary-Lemon points out, this sense of ending is even built into the rhetoric of the "anthropocene" that suffuses mass extinction discourse: "creating a geological epoch named after ourselves suggests the impossibility of imagining our way through it—after all, the end of the *anthropos* suggests the end of human life on the earth."[13] Despite the ubiquity of apocalyptic rhetoric in environmental discourse, some critics have suggested that it is self-defeating and may lead to skepticism or fatalism, while others argue that it may serve as a moral catalyst for political action.[14] In this chapter, I offer another way of thinking about human extinction beyond the apocalyptic frame. Using a comparative analysis of two social movements that position human extinction in relation to the mass extinction crisis, I show how the prospect of extinction presents an ideal case with which to examine the rhetoric of "species thinking" in vital advocacy.[15] Species thinking emphasizes the "basic sameness" of humanity, as Paul Gilroy terms it, and focuses on a sense of shared "vulnerability and finitude" in the face of environmental and medical crises on a planetary scale, such as climate change and pandemics.[16] "All calls for species thinking," writes Shital Pravinchandra, are anchored by a sense of biological similarity, while their ethical power is underwritten by a "rhetoric of life preservation."[17]

The first group I examine, the Voluntary Human Extinction Movement (VHEMT), argues that the extinction crisis can be solved by gradually phasing out the human species through deliberate nonprocreation—the "inhuman" proposal of Peter Reed that I discussed in chapter 2. VHEMT was founded in the US in 1991 and loosely organized through a newsletter and website. Participation in it has waxed and waned over the years. However, it has recently gained new life in digital forums such as Facebook and Reddit.[18] While there are clear overlaps between advocates of human population control and VHEMT, there is one notable difference: population control advocates argue for the reduction of human populations for the sake of human (and sometimes nonhuman) lives. VHEMT argues for the elimination of human life for the good of life itself. The second group, Extinction Rebellion (XR), was formed in the UK in 2018 to "defend life itself" through nonviolent direct action.[19] While climate change is a major focus of the group, XR is strikingly different from other climate change activism in that it frames the climate crisis *as* a crisis of mass extinction that includes potential human extinction (unlike VHEMT, XR sees human extinction as a bad thing). XR charges that governments have broken their contract with their constituents by ignoring and perpetuating these ecological crises and demands that the authority for solving them be given back to everyday people.

Jeffrey Nealon suggests that the "external night of extinction," and particularly the prospect of *human* extinction, may prompt a "radical" shift in "our understanding and practices of life."[20] In my more psychoanalytic moments, I wonder if mass extinction is the terrifying engine of the vital turn that Robert Mitchell identifies in the present moment. But whatever the thrumming existential anxieties it may be fueling, the threat of mass extinction provides the exigence for humans, as Ursula Heise argues, to shift our sense of ourselves "as a species among species," united in life itself through the threat of death itself.[21] VHEMT and XR both make extinction an object of political action, but where they differ is in the way they rhetorically position the human species in relation to other forms of life. In what follows, I look closely at these two groups using a concept I call *bioplurality*. Drawing from Hannah Arendt's concept of plurality with a modification inspired by Sylvia Wynter's critique of species thinking, bioplurality names a "twofold character" of life that acknowledges similarity and difference across *and* within species.[22] This change in self-conception involves an understanding of humanity that is both distinct from (species) *and* connected to other forms of life on Earth (among species). But while a powerful way to imagine humanity-as-collective-agent *and* patient during this moment of crisis, species thinking is not without its problems.[23] In its focus on commonality, species thinking risks overlooking the crucial differences by which we can locate responsibility and the effects of power, as well as the "new forms of exploitation in the terrain of life itself" that Pravinchandra argues that species thinking can give rise to.[24] Species thinking thus requires a careful attention to plurality—not just plurality among species but plurality within our species as well.[25]

Bioplurality

Plurality is the cornerstone of Hannah Arendt's theory of political action in *The Human Condition*, yet it is also remarkably undertheorized there. This may be because for Arendt, plurality is not a theory but a given: it describes "the *fact* that men, not Man, live on the earth and inhabit the world."[26] Plurality names the "twofold character" of the human condition: by virtue of being born into the human species, "we are all the same, that is human," and yet we are all different; we are all human "in such a way that nobody is ever the same as anyone else who has ever lived, lives, or will live."[27] Plurality thus identifies two fundamental elements—equality and distinction—of human being that must be held in mind at the same time. "If humans were not equal," writes Seyla Benhabib, "they could not understand each other; if they were not distinct, they would need neither speech nor action to distinguish them from each other."[28] This realm of speech and action is what Arendt calls the "world," the place where human beings transcend the limits of biological life to become fully human as such, transforming from *whats* into *whos*.[29] The world "is not identical with the earth or with nature, as the limited space for the movement of men and the general condition for organic life," Arendt writes; rather, the world is a human artifact, created through speech, action, and the presence of others who give it meaning.[30]

Considering this chapter's focus on life and death, it might seem odd that I take inspiration from Arendt, particularly *The Human Condition*, which is sometimes characterized as an "anti-biological" project.[31] In Arendt's normative vision of politics, she famously wrote, "not life but the world is at stake."[32] However, there is more liveliness in Arendt's work than it would first appear. Whether through her discussion of Sputnik's escape of Earth or her concerns about life extension and artificial reproduction, Arendt positions her intervention in *The Human Condition* in response to a timeless desire to transcend the animal limits of biological life. Arendt links these emerging technologies of life with the emergent technologies of death; writing just a few years after the development and deployment of nuclear weapons, Arendt warns that there is "no reason to doubt our present ability to destroy all organic life on the earth."[33] Technoscientific control over life and death—extending all the way to mass extinction through nuclear annihilation—is not just a political question, Arendt explains, but a "political question *of the first order*."[34]

This brief line is telling. It suggests that the different aspects of Arendt's *vita activa*—labor, work, and action—are not separate spheres, as they are sometimes characterized, but layers built one on the other.[35] If Arendt worries that the concerns of labor and work, gathered together as social concerns, threaten the political, it is because they are—indeed, must be—taken for granted, in the same way that buildings are built on stable ground. "Ground" here is *literal* ground. "The earth is the very quintessence of the human condition," Arendt writes, "and earthly nature, for all we know, may be unique in the universe in providing human beings with a habitat in which they can move and breathe without effort and without artifice."[36] Life itself and death itself are political questions of the "first order," then, not because they are more definitive or more properly political than others but because life, and the planet that makes it possible, is the foundation on which the human world depends.[37]

I haven't the space to detail the many critiques of Arendt's central theses in *The Human Condition*, particularly since her writing on plurality is not the location of this chapter but its place of departure. But one critique is directly related to my argument: the gaps and frictions in her work concerning identity and difference.[38] Earlier, I suggested that plurality is not extensively theorized in *The Human Condition* because it is presented as a matter of fact. Yet the "Man" at the heart of Arendt's plurality emerges not from a womb but from a story. As Sylvia Wynter argues in her essay "Unsettling the Coloniality of Being/Power/Truth/Freedom," extending a line of anticolonial thinking with roots in Frantz Fanon and Aimé Césaire, "Man" is the outcome of an effort to present a dominating subset of humanity as human *being*, the unmarked, universal center of *Homo sapiens*, the referent that depends on the exclusion of human Others for its legibility and existence.[39] Man, in other words, has been the rhetorical, epistemological, and ontological accomplishment of white, Western, Enlightenment elites. In Wynter's tracing of Man's historical emergence under the conditions of theology, colonialism, modernity, and racism, she offers a powerful reminder that appeals to species commonality in Man's name are difficult to unravel from the discourses of hierarchy and violent forms of domination and exclusion that gave him birth. It is a mistake to say, however, that Wynter is opposed to speaking of humanity as such, or what she calls an "ecumenical" humanity.[40] Rather, she is opposed to the overrepresentation of the "West's biocentric Man" in our stories of human origins *and* human futures, "the incorporation of all forms of human being into a single homogenized descriptive statement that is based on the figure of the West's liberal monohumanist Man." The problem with this figure, "ostensibly natural-scientific," she explains, is that it presumes that the human is "like all purely biological species, a natural organism, . . . and therefore exists, as such, in a relationship of pure continuity with all other living beings (rather than in one of *both continuity and discontinuity*)."[41]

Borrowing from Arendt's concept of plurality, then, and with attention to Wynter's critique of the species thinking at its center, what I am calling "bioplurality" names a "twofold character" of living organisms that acknowledges similarity and difference across *and* within species and holds this tension in mind as a point of generative dissonance. "Bioplurality" names a way to understand the human connection with the other-than-human world that acknowledges both the similarity and connections we have with other living beings—continuity, as Wynter describes it, but also discontinuity: the difference and distinctions with which we might identify not only our different interests but also the differential effects of power.

"The Gift of Life": The Voluntary Human Extinction Movement

The Voluntary Human Extinction Movement is the life's work of Les U. Knight, the pseudonym of an Oregon environmentalist, veteran, and substitute teacher. Knight prefers not to be called the "founder" of VHEMT: he claims that he just provided a name for ideas that have been circulating for centuries. VHEMT launched in 1991 as a photocopied newsletter titled *These Exit Times*. The newsletter shifted online in 1996, and VHEMT now exists primarily through its website, where it is described as "a movement advanced by people who care about life on planet Earth." As its name makes clear, the goal of VHEMT is the voluntary reduction of the human population to the point where the species goes extinct, allowing the Earth (often called "Gaia" in VHEMT's materials) to "heal" from human damage and leaving the rest of the living world "free to live, die, evolve" without our meddlesome, destructive presence.[42]

While the objective of VHEMT has led critics to see its members—called "Volunteers"—as "misanthropes and anti-social, Malthusian misfits," Knight insists that the aim of the group is a "humanitarian" one: fewer humans on the planet would result in less suffering for humans and nonhumans alike.[43] While VHEMT sometimes describes Earth's "illness" as a general kind of environmental degradation, it is clear that the primary, most urgent problem it seeks to solve is mass extinction. "The hopeful alternative to the extinction of millions, probably billions, of species of plants and animals is the voluntary extinction of one species; Homo sapiens . . . us."[44] For VHEMT, humanity is the quintessential *pharmakon* for Earth's poor health: its disease as well as its cure.

Birth control is the primary means by which VHEMT aims to achieve its objective, and sterilization through vasectomy is represented in especially glowing terms. Vasectomy is described as "tying the lover's knot," a "gift" that the fertile, heterosexual male Volunteer can give his wife or girlfriend, a "mission of duty," and a "divine act." VHEMT frequently presents vasectomy as a feminist action in that it allows men to take responsibility for contraception and thus frees women from the effects of birth control and unwanted pregnancies. One Volunteer writes wistfully of a future in which all male children receive vasectomies as an initiation into manhood: "a population of sterile men would not dominate women through their genetic claim on women's offspring."[45] But the gift goes further than that.

Vasectomy—and nonprocreation more generally—is not just a gift to one's lover but a show of commitment to life itself. "By allowing wildlife to take the space which their additional descendants would have taken," Knight explains, "these Volunteers are giving the gift of life to planet Earth."[46] For VHEMT, life itself appears to be a zero-sum game between human and nonhuman worlds, with echoes of Linkola's rhetoric from chapter 2. If *their* life has been threatened by *our* life, this argument suggests, then *their* life is made possible by *our* death. This position is motivated not by a hatred of humanity, VHEMT insists, but by the "natural abundance of love and logic within each one of us."[47] One Volunteer explains, "People need to understand that we (humankind) are not stewards of Earth or our fellow Earthlings. If anything, we're Nature's children, . . . and as such we can save our entire family (Well, what's left of it) by taking this opportunity to bow out while there's still time."[48] As self-appointed caretakers of Earth, humans have failed. By reidentifying with the "natural family" and by denying an instinct to reproduce, this argument goes, humans might be able to atone for the terrible harm we've caused.

Continuing this familial narrative, *These Exit Times* features numerous instances when nonhuman animals are described as substitutes for the children that Volunteers choose not to have. If humans feel the need to nurture, Knight suggests that caring for "Earth's 'children' can be a viable alternative. Wildlife rehabilitation and protection, habitat preservation, reforestation, Adopt-a-Stream, and gardening are some possibilities." As one Volunteer explains, "Every time I see a turtle sunning itself, or a deer bouncing across my path, I will say, 'there is my child.'"[49] One woman writes that the space in her home that would be devoted to a baby "is now home to another kind of the world's homeless, animals that were thrown away like yesterday's trash": "I consider them all my children, a bit furry maybe, but they are still a part of me."[50] Continuing this theme explicitly are two comics in which a woman chooses to give birth to a bonobo, "our closest relative," rather than a human child. "Ahh . . . ," the woman sighs, "guilt free pregnancy."[51]

Here, we see bioidentification motivating two different kinship roles. As "nature's child," one acts out of filial piety, a duty to Gaia, the Mother Earth who gives us life. But far more frequent are discussions of humanity as nature's parent. The refusal to procreate is described not in deathly terms but in lively ones—nonprocreation helps to create new life by "making space" for it and also by nurturing it in a variety of ways. However, by placing humanity in the role of nature's parent, that is, by placing humanity in Gaia's role, VHEMT ironically places humans *outside* the living world, vulnerable to no threat, beholden to no living creatures save those we deem worthy enough to take in as our wards.

Knight does not call VHEMT an organization, preferring the word "movement," which "refers to the transfer of an idea from one person to another."[52] The primary objective of VHEMT as a *movement*, then, is an evangelical one. It is not a belief system or a code of behavior but a "concept," and the goal of Volunteers, beyond not reproducing, is to spread the word to others.[53] By 1994, Knight claims that publications large and small had been reporting on the movement and that VHEMT Volunteers had been sharing their message at environmental gatherings, at conferences, on radio shows, and anywhere they had the chance, to positive effect. (I first became aware of the group after encountering its newsletter in independent record stores in the mid-1990s and seeing *These Exit Times* reviewed, with some measure of amusement, in punk fanzines.) As a result of the circulation of *These Exit Times* and its message, Knight notes in 1994, "awareness is rising and The Movement is advancing."[54]

To help attract others to VHEMT's cause, Knight lays out a number of explicitly rhetorical strategies in *These Exit Times* for his readers to use in order to persuade potential Volunteers into the fold. Volunteers are urged to keep their overall message positive and to focus on "good things" rather than doom and gloom (presumably a risk when trying to persuade people—one cannot overstate this—*to support human extinction*). Knight calls for an audience-centered approach and argues that Volunteers ought to tailor their messages accordingly. Sometimes the best appeal is a "personal" one. Knight suggests, "Setting an example by living well and being happy presents our decision in a good light.... Never underestimate your passive influence."[55] Likewise, Volunteers are urged to respect the "magical" power of the words they use to describe the movement. Euphemisms, Knight explains, "may seem deceitful or even cowardly, but, unless we're preaching to the converted, word choice can make a magically powerful difference in how well the VHEMT view is perceived." Knight thus urges Volunteers to avoid using the terms "population control," "birth control," or "sterilization," which sound "like the curse of fascism."[56]

As the direction on word choice reveals, Knight and other VHEMT Volunteers are well aware of the dark history associated with population control, which, as I discussed in chapter 2, has long haunted the edges of Western environmentalism. They thus take pains to distinguish their motives as *anti*eugenic, *anti*racist, and *pro*feminist and explicitly state—more than once—that they are not in favor of famine, plagues, or killing of any kind. Discussing the "stop at two" recommendation of the group Zero Population Growth (ZPG), for example, Knight writes that this message "actually *encourages* reproduction by 'qualified' couples"—an implicit critique of ZPG as eugenicist.[57] Later in the issue, the topic of eugenics is taken up more explicitly. Directly taking on the arguments that proposals by VHEMT will result in the "wrong people" reproducing, Knight writes that these arguments contain the "attitude that some people are the right people to pass along their genes": "The mindset behind this bloodline mentality is deep and strong: more of 'us' and less of 'Them.' Smells like racism to me. When couples try to conceive a specific gender, sexism is also in the wind. It goes beyond racism for us to recreate replicas of ourselves while tens of thousands of Others' children die from lack of care each day.... Breeding for power is a remnant of that ancient tradition of mass murder we call genocide. The motivation remains the same."[58] Still, while VHEMT appears to offer an approach to population control in which racial, socioeconomic, or geographic factors do not play a role, a closer look reveals a much more complicated story.

For example, a letter from one Volunteer to *These Exit Times* expresses concern about "the declining birth rate of the educated and/or wealthy classes," who are precisely those "enlightened few . . . who would adopt this philosophy/practice." Yet "the ones you may never reach are the very ones most responsible for overpopulation of the planet"—specifically, "the 3rd world." The letter writer then asks why economic incentives are seen by VHEMT as a form of genocide. Knight responds, "Because discrimination against racial and ethnic groups has caused economic inequities, the long-term result of paying people to not reproduce could be a subtle form of genocide. *It's worth the risk, though*, . . . [for] promoting procreation and denying reproductive rights, especially when famine is rampant, causes more suffering than plain old-fashioned genocide."[59] Hence, VHEMT claims to offer a solution to racist eugenics by substituting a kind of pro-life, color-blind antihumanism in its place. It's not that the "wrong people" are reproducing, Knight insists; it's that the wrong *species* is reproducing: "Regardless of our *superficial* differences, we are all Homo sapiens."[60] Rather than getting "past" race by emphasizing positive similarity over negative difference, the form that color-blind racism often takes, VHEMT flips the equation in what amounts to a tidy syllogism: humans are guilty of capital crimes against the living world; we are all human; therefore, we all deserve to die.[61]

Though all humans are implicated in this argument, Knight is adamant that VHEMT does not favor forms of "involuntary extinction."[62] To emphasize this point, Knight features an interview with "Geophilus," a representative from the Gaia Liberation Front (GLF), an extremist group associated with the Church of Euthanasia (slogan: "Save the Planet, Kill Yourself") that also wishes to see the human species eradicated.[63] Knight includes the interview in what appears to be a clear attempt to make a distinction between two positions: the GLF has a grim view of humanity, while VHEMT finds hope in humanity's inherent "love and logic" that would allow us to gracefully bow out for the good of life on Earth. "The Humans evolved *on* the Earth," Geophilus explains, "but are no longer *of* the Earth. Having become alienated from the Earth, they must be regarded as an alien species." Accordingly, the GLF's position on human extinction "does not exclude the involuntary route," such as forced sterilization, the release of an airborne virus, or even genocide. "If you want the Humans to die out," Geophilus reasons, in an echo of the ecofascist rhetoric I discussed in chapter 2, "is it so awful if some of them die out before the rest?"[64]

Despite VHEMT's protestations to the contrary, what draws VHEMT in line with ecofascists like the GLF and Pentti Linkola is an antihumanist form of species thinking that situates human life in an inverse relationship to the rest of life on Earth. Simply put, when VHEMT speaks of life itself, humans are not imagined to be a part of it, except as an all-powerful agent on a passive living world, a reproduction of the "institutionalized, long dominant Euro-Western fantasy that all that is fully human is fallen from Eden, separated from the mother, in the domain of the artificial, deracinated, alienated, and therefore free."[65] Though given lip service as a "child of nature," humans are still able to step outside its constraints, thus "enacting the same hubris that results in dispositions toward non-human nature" at the root of the problem it seeks to solve.[66]

While there is no question that humans—in particular, those from the wealthy West and Global North—are the cause of the current biodiversity crisis, the extinction of any species extinguishes multiple links and connections in the living world, reverberating through its ecosystem in ways that cannot be anticipated in advance. I am reminded on this point of Alan Weisman's thought experiment in human vanishing, *The World Without Us*, in which he wonders, "Is it possible that, instead of heaving a huge biological sigh of relief, the world without us would miss us?"[67] We live with and because of other living creatures, but many also live with and because of us. Not only does VHEMT downplay the differences within our species at the root of the current extinction crisis, in other words, but by severing the connections by which humans are joined to other species and they are joined to us, it also misses how our lives, our deaths, and our futures are intertwined.

"We Are Life": Extinction Rebellion

On October 8, 2018, the UN's International Panel on Climate Change (IPCC) released a landmark report on the state of anthropogenic global warming. In the report, the panel (famously staid in its tone and conservative in its predictions) declared that the world had just twelve years to rapidly decarbonize in order to keep global warming to 1.5° C, the target set by the 2016 Paris Agreement. According to Debra Roberts, cochair of the IPPC impacts working group, the report presented the world and its leaders with "a line in the sand and what it says to our species is that this is the moment and we must act now."[68] Shortly after the October report, Extinction Rebellion (XR) published two open letters in the *Guardian* signed by hundreds of academics and environmental activists calling for immediate action, and the group officially launched on October 31, 2018.

Notably, XR's inaugural letter does not mention climate change, except obliquely. Rather, it sets biodiversity loss as what Lloyd Bitzer would call the "controlling" exigency of the movement.[69] "We are in the midst of the sixth mass extinction," the letter reads. "If we continue on our current path, the future for our species is bleak."[70] Species thinking is prominent in XR's discourse; however, as the letter suggests, rather than succumb to the antihuman exceptionalism that characterizes VHEMT, it positions humanity in relation with the rest of the living world—a species among species. Humans are agents and acted-upon, perpetrators and victims. We are joined with the rest of the living world through the common substance of life and thus also by the looming threat of death itself.

XR has a remarkably consistent message, which unifies the movement even as hundreds of individual chapters have popped up around the world only a few years after XR's founding. Simply put, XR has a *brand*—a free-to-download, colorful, and punk-styled brand, perhaps, but a brand nonetheless, complete with its own font, logos (hourglass, bee, skull), and slogans ("Rebel for Life"). As a result of the organization's branding, XR's discourse and imagery is recognizably XR, including its demands, which are presented as easy-to-remember slogans:

- Tell the Truth: Government must tell the truth by declaring a climate and ecological emergency, working with other institutions to communicate the urgency for change.
- Act Now: Government must act now to halt biodiversity loss and reduce greenhouse gas emissions to net zero by 2025.
- Beyond Politics: Government must create and be led by the decisions of a Citizens' Assembly on climate and ecological justice.[71]

As the demands make clear, XR brings the climate crisis and extinction crisis together, but it does not position one above the other, connect them via simple cause and effect, or collapse them into a single "ecological crisis." One can almost see the pencil sketches of the discussions crafting the demands in order to keep biodiversity at the forefront of the movement.

While the demands make clear that halting extinction is a concrete objective of the group, it also serves a framing function. That is, XR presents climate change *as* an issue of mass extinction. As I mentioned earlier, biodiversity loss gets far less press attention than climate change does, which makes it an odd choice of frame, if the point is to inflame public concern and inspire political action. But this is where XR's emphasis on *human* extinction comes to matter a great deal. According to Gail Bradbrook, one of XR's founders, the focus on human extinction was a rhetorical choice: "We think it is important to talk about the possibility of human extinction in order to expand the window of acceptable discourse on climate change and ecological collapse but we also acknowledge that this is not only about our species and that the web of life is intricately interconnected."[72] Act now, the second demand implies, or *we all die*. The "we" here under threat of extinction is rhetorically identified not by nation, not by species, not even by bioregion or ecosystem but, again, by the "web of life."

On April 15, 2019, XR launched "Rebellion Week," a concurrent set of nonviolent direct actions in over eighty cities in thirty-three countries across the world. In the epicenter of London, thousands of Rebels blockaded Oxford Circus, Piccadilly Circus, Waterloo Bridge, and Parliament Square. At Marble Arch, XR activists dropped a large pink sailboat—christened the *Berta Cárceras* to honor the murdered Indigenous Honduran activist—in the intersection. On the ship's hull was written XR's first demand: "TELL THE TRUTH." Rebels smashed the revolving door at the British headquarters of Shell Oil, holding signs that read, "Wanted for Ecocide," and others glued themselves to a train at Canary Wharf. More than a thousand people were arrested during Rebellion Week, including Farhana Yamin, an environmental lawyer who has represented small island states at the UN and who helped to draft the 2016 Paris Agreement. Other XR chapters held similar, though smaller, actions across the globe. In Portugal, protestors hoisted giant bouquets of empty plastic water bottles in front of Nestlé headquarters. In South Australia, "grandparent" activists occupied the House of Assembly. In Sweden, activists held a die-in in front of the Swedish parliament. In Berlin, XR members blocked Oberbaum Bridge. In New York City, under a large banner that read, "DECLARE CLIMATE EMERGENCY," activists blocked traffic with a die-in on the streets in front of City Hall, which resulted in sixty arrests.

If the signs and chants are any indication, climate change was the main focus of Rebellion Week, but the extinction theme remained prominent. Many Rebels dressed as animals or wore insect wings (a popular choice with the many children who showed up to the events). Dinosaur imagery was common—one London activist's sign read, "The Dinosaurs Didn't See It Coming, Either." One Rebel, dressed as a dodo, warned, "YOU'RE NEXT HUMAN." In London and in Paris, hundreds of Rebels gathered at natural history museums and lay down in a massive die-in in the main halls under skeletons of dinosaurs and a great blue whale, while others handed out literature on the sixth mass extinction to onlookers. In these orchestrated "image events," activists' living bodies, arranged under the massive skeletons of extinct species, make the argument that not only is mass extinction *not* history but it is happening now, and humans are not exempt.[73] Here it is not life itself but death itself that identifies humanity with the rest of the living world; the dinosaurs and dodos are ghostly heralds of what is coming for all of us.

A recurrent theme in XR discourse is extinction as a multispecies risk, and a common method of visually rendering this idea is to juxtapose human skeletons with the skeletons of other animals, such as the image found on one of the posters available for download on XR's website. This poster has a simple design: the word "EXTINCTION" on a bright pink background and a collection of skulls of a variety of species. Underneath, the phrase "everyone gone forever" brings all species together using a pronoun typically applied only to humans. The effect does not anthropomorphize these species so much as it emphasizes commonality through shared threat: the web of life that is, at the same time, a web of death. The human skull is not at the center (no skull is) but on the periphery, a species among species.

There are two human skulls in the image: one an adult's, one a child's. The adult skull is a recognizable image—the iconic symbol of threats to human life—but the child's small skull is unfamiliar to the point that it might be missed by the casual viewer. Once seen, however, it is a shocking reminder that the category "human" includes children as well as adults. In one way, this injects a small measure of human difference into the image, but it primarily serves to encourage a sense of species thinking across time. Children (and grandchildren) are often invoked in climate change discourse as a trope for the future. But it is rare to see the death of children invoked in such a literal way. One act of civil disobedience along these lines is to dump buckets of "blood" in public spaces, a tactic that has been used by several XR chapters. One blood-spilling action in

London on March 12, 2019, was called "Our Children's Blood." "The spilling of blood is symbolic of the risk to our children's futures," explains one Rebel on a video that XR produced of the action. "My own children could be looking at being killed for a loaf of bread because there's no food around." "It's also about all the other species that are on the brink of extinction," adds a young boy. As activists, including young children, dip their hands and leave bloody prints on the walls of 20 Downing Street, a young girl at the microphone tells the crowd, "We don't want this blood to be ours or theirs, so keep it up. Tell the truth, and rebel for life."[74] During Rebellion Week, teenaged XR rebels staged a protest at London's Heathrow Airport holding a banner that read, "ARE WE THE LAST GENERATION?"

As the young Rebels' message suggests, XR brings together not only the human and other-than-human world but also the past, present, and future in an affective mix of warning and mourning.[75] Some of the most vivid examples of this entanglement are the multispecies funeral processions that have been a feature of many XR events. During Rebellion Week, Rebels occupying the area near Westminster Abbey hoisted life-sized "skeletons" of rhinoceroses, gorillas, and humans as they marched behind a large black banner proclaiming, "Life or Death," followed by a brass band playing a funeral dirge. The activists marched in a slow procession in the spirit of a New Orleans jazz funeral, complete with umbrellas emblazoned with XR's hourglass icon. Near the funeral march, activists created a makeshift graveyard in a nearby green space with painted cardboard headstones emblazoned with species lost to extinction and the dates they disappeared. One headstone featured the outline of a human figure and, below it, a question mark. In New York City during Rebellion Week, XR activists held an event called "Extinction Mass: In Remembrance of Lost Species," in which participants—asked to wear funereal black or animal masks—were invited to gather, march, and "mourn our lost relatives."[76] Participants carried a makeshift coffin painted black with the message, "OUR FUTURE," the "our" here clearly indicating both human and nonhuman futures. While some people may see grief and mourning as acts of the past or as indicative of antipolitical despair, Thom van Dooren argues that these responses instead form a "foundation of any sustainable and informed response" to the extinction crisis.[77] XR activists often speak of grief in this way, describing coming to terms with one's grief as the necessary first step for effective political action.[78]

Judith Butler argues that one of the marks of life *as* a life, that is, a life with value, is grief.[79] Grief for the other-than-human world helps us to understand how our living and our dying is "entangled," and it serves, as Joshua Trey Barnett writes, as an explicit "affirmation of our interconnectedness with the earth and our cohabitants on earth."[80] "We are something other than 'autonomous' in such a condition," Butler writes, "but that does not mean that we are merged or without boundaries."[81] Even as we identify with our human and other-than-human relations through the common life, vulnerability, and finitude revealed by the threat of mass extinction, in other words, it is important to remember that we are not identical.

While XR has been quick to emphasize the connections between humans and other species, it has been slower to recognize the differences within our own. Shortly after Rebellion Week, a coalition of activist groups calling themselves "The Wretched of the Earth," in a nod to Frantz Fanon, called on XR organizers to reconsider the role of difference in the organization going forward in its composition, tactics, and demands. If you recall, XR's first open letter predicts that unless radical action is taken, "the future of our species is bleak." But for many people around the world, the letter explains, especially Indigenous communities, communities of color, and those living in the Global South, a bleak ecological future is a "birth right." "Our house," the letter states, referencing Greta Thunberg's 2019 speech in Davos, "has been on fire for a long time."[82] As several scholars have argued, the fears of future apocalypse found in much Euro-Western environmental discourse do not take into account the experiences of Indigenous people, many of whom describe living in an apocalyptic *present* or what "our ancestors would have understood as a dystopian future."[83] XR's species thinking focuses on the threat of human extinction, which by definition has not happened yet. As such, it overlooks how many human and nonhuman worlds have ended already, how many are ending right now, and how many have endured, even despite *intentional* efforts at extinction.[84]

As critics of the term "anthropocene" have argued, the rhetorical act of flattening humanity into *anthropos*—an invocation of Wynter's Man by another name—occludes the radically different relationship that humans have to the climate crisis and extinction crisis, with regard to both responsibility for those crises and their impacts in the past, present, and future.[85] Nowhere is this difference more keenly felt than in the "cruel irony" of climate change: those who live in areas that have contributed the least to the problem are those who are being impacted first and will suffer the most, with Indigenous peoples most acutely at risk.[86] However, this vulnerability is not just the result of present-day threats from sea-level rise, Kyle Whyte argues, but also an effect of the past: namely, the relocation of many tribes from expansive lands to small parcels with "limited adaptive options."[87] The prospect of human extinction allows—I would hope, demands—humans to identify with each other and with other forms of life against the threat of death itself and helps us to imagine how our futures are bound together, a point I will return to in this book's conclusion. However, as the Wretched of the Earth make clear, species thinking also blocks from view other "biological annihilations" in the past and present and the violence, both agonizingly slow and horrifyingly quick, that has marked, and continues to mark, some human lives as more "life" than others.

On May 1, 2019, just one week after the last activist left Central London, the British Parliament unanimously passed a declaration of climate emergency, fulfilling XR's second demand. While heartening in many respects, the declaration is nonbinding and is not tied to any funding or specific policy. More disheartening is the fact that the declaration is almost entirely about climate change—the biodiversity crisis merits only a brief mention in the middle. The main point of the declaration, in other words, appears to be a call to protect the Earth to ensure a future for *human* life.

Later the same month, another landmark ecological emergency was declared by the Vuntut Gwitchin First Nation of the North Yukon, the first such declaration by an Indigenous nation.[88] The declaration is titled "Yeendoo Diinehdoo Ji'heezrit Nits'oo Ts'o' Nan He'aa," which means "After Our Time, How Will the World Be."[89] The declaration states that the Vuntut Gwitchin people have a duty "to the past, present, and future generations of all living beings" and affirms about their culture, "[it] teaches us how to live with this world in a way that provides us with healthy animals, lands, waters, ecosystems, and populations." The declaration references the climate crisis as a planetary one and notes the specific vulnerability of other Indigenous people around the planet, but it also situates its exigency in the Vuntut Gwitchin's specific experience in their arctic lands, an area warming at twice the rate as the rest of the planet.[90] Affirming a commitment to the Paris Agreement to keep warming below 1.5° C, the declaration also centers Indigenous knowledge and leadership, calling for an Indigenous Climate Accord to coordinate efforts with their "relatives around the world."

Nowhere does the declaration use the word "extinction," nor should we expect it to. The declaration does not express a concern for species or a defense of an abstract "life itself"; rather, it is a call to action against a threat to the lives of the Vuntut Gwitchin people present and future, their livelihood, their way of life, and the lands, waters, and animals to which they are related. Threats to the caribou are especially grave. The Gwitchin, explains spokesperson Sarah James, "are caribou people": "Caribou are not just what we eat; they are who we are. They are in our stories and songs and the whole way we see the world. Caribou are our life. Without caribou we wouldn't exist."[91] James's description of her people as "caribou people" shows how inadequate words like "connected" or "identification" would be to describe this relationship—and, in an echo of my argument in chapter 1, it also reveals the limits of bioplurality, which, though it draws on anticolonial thought, is still grounded in a Western understanding of the relationship of human and other-than-human worlds. The Gwitchin are not *connected* to the caribou, because they are not separate from them; they do not *identify* with the caribou because of a common substance, life or otherwise, because they were not distinct to begin with. *They are who we are. Without the Caribou, we do not exist.*

For those of us not taught to see the other-than-human world as kin, it may be difficult to understand the complex moral bonds and mutual responsibility that are reflected in the Gwitchin statement. But if meaningful political action is to be taken at this critical moment, a different way of thinking is essential. Species thinking is a useful tool to imagine how human and other-than-human lives, deaths, and futures are intertwined, and it serves an important rhetorical role with regard to scholarship, advocacy, and policy. But species works like any terministic screen. By focusing tightly on similarity, species thinking occludes the differences by which we may identify the working of power: the fulcrum of both resistance and justice. In contrast, this chapter has offered the idea of bioplurality, a way of thinking with, within, and beyond species, toggling back and forth between individual and kind, similarity and difference, connections and distinctions, noting resonances while also listening carefully for moments of dissonance. By highlighting similarity, bioplurality avoids the twin traps of humanist and antihumanist exceptionalism that assume that humans are outside the living world or that we are somehow exempt from threats to it. By attending to difference, bioplurality marks humanity's unique culpability for the current crises but also its unequal responsibility for those crises and their unequal effects.

Coda: The Fourth Demand

When I first started writing the article on which this chapter is based, XR had a very limited presence in the United States. During Rebellion Week of April

2019, I attended a protest by the Chicago XR chapter, which at the time consisted of about three dozen people. Stimulated by XR's success in the UK and the staggering numbers of youth climate activists on the streets each Friday in the early months of 2019, XR-US underwent a rapid expansion. In October 2019, I attended another Chicago action that drew well over a thousand XR members and allies from other environmental and social justice movements. After a festival-type gathering that lasted a few hours, which featured food, dancing, art, and spoken-word performances, activists took to the streets for hours, shutting down a good portion of downtown Chicago well into the late hours of the evening (fig. 2). While I attended the first event as a researcher, I attended the second as an activist. Inspired by XR's philosophy of nonviolent civil disobedience and wanting to channel my despair about the ecological crisis into some kind of good, I joined my local XR group in the summer of 2019. In that group, I found camaraderie with people who had a fierce passion for environmental and

Fig. 2 | Extinction Rebellion activists in downtown Chicago, October 7, 2019. Photograph by the author.

social justice, many of whom were willing to risk their freedom to bring public attention to the ecological crisis. I participated in several actions in the fall of 2019, also serving as our local Media and Messaging team leader until a serious illness—first my own, then the world's—required that I step away. Because I was with them as an activist and not a researcher, I will not write about those experiences except to thank my friends in XR for the community and inspiration they offered me during a very dark time.

Although I was inspired by the success of Rebellion Week in the UK, I also agreed with critics of the organization, particularly its many lacunae around issues of race and privilege. What encouraged me to join XR, however, was an early decision by the national leadership of XR-US to add a fourth demand to the UK's original three. This demand reads, "We demand a just transition that prioritizes the most vulnerable people and indigenous sovereignty; establishes reparations and remediation led by and for Black people, Indigenous people, people of color and poor communities for years of environmental injustice, establishes legal rights for ecosystems to thrive and regenerate in perpetuity, and repairs the effects of ongoing ecocide to prevent extinction of human and all species, in order to maintain a liveable, just planet for all."[92] As Bea Ruíz, a member of the XR-US's national team who helped to craft the demands, explained the rationale, "that's the only thing that's going to work, and it's the only morally right thing to do."[93] While addressing the issue of human extinction, XR-US's fourth demand is precisely the kind of bioplurality that I have described in this chapter—an acknowledgment of similarity and difference with an attention to the differential effects of power. Even further, the demand locates the unfolding ecological crisis not only as a present and future threat but also as an effect of the past, an accumulation of injustice over many years. Several global chapters have adopted XR-US's fourth demand, and a distinct chapter, a coalition of both XR- and non-XR-affiliated activists, has emerged in line with its thinking, Global Justice Rebellion, which seeks to put the people on the front line of the environmental crisis at the center of XR's demands, activism, events, and leadership. Critiques of XR from groups like the Wretched of the Earth, as well as from members within the movement, have resulted in critical conversations in the organization about XR's tactics (specifically, the use of voluntary mass arrest, which, though not required by members, is enabled by racial and economic privilege) as well as its strategies, objectives, and understanding of what makes the ecological crisis a problem in the first place.

But the growing attention to difference within XR also provoked division.[94] In the spring of 2020, several US activists formally split with the XR-US, renaming themselves "XR America." Their specific point of contention was the fourth demand. According to one XR America member, "If we don't solve climate change, Black lives don't matter. If we don't solve climate change now, LGBTQ [people] don't matter. If we don't solve climate change right now, all of us together in one big group, the #MeToo movement doesn't matter. . . . I can't say it hard enough. We don't have time to argue about social justice."[95] Arguing that XR-US's focus on frontline communities and their commitment to social justice created delay and division, the new group removed the fourth demand, initially replacing it on its website with the slogan "one people, one planet, one future."[96] By abandoning the fourth demand, and its call for responsibility and repair, XR America engages in species thinking that *intentionally* flattens difference—and in so doing, it ensures that the "one future" it seeks will be no different than the past.

In *Out of the Dark Night*, Achille Mbembe writes that "abstraction of differences is not a condition sine qua non for consciousness of belonging to a common humanity." However, if we want to "open the future to everyone," it is first necessary to understand the conditions under which the relations of subjection were first established and reproduced, "a task that must go hand in hand with a critique of all forms of universalism that, hostile to difference, and, by extension, to the figure of the Other, attribute the monopoly on truth, 'civilization,' and the human to the West."[97] As Mbembe argues, attention to difference does not preclude speaking of a common humanity or of a life in common with other Earthly creatures. One can be in a universal predicament and identified with others through the threat of that predicament, while remaining aware of how differences in power, unfolding across history, produce different vulnerabilities, culpabilities, and responsibilities for repair. Mass extinction threatens all species, yes, but not in the same way, at the same time, or at the same intensity. Climate change threatens all humans, yes, but not in the same way, at the same time, or at the same intensity. The task, then, is not to *preserve* "our" life and protect "our" future as it has been unfolding but to find the route to a new future and the new us that the journey will reveal.

4

"This Universe Belongs to Life" | Planetary Protection and Planetary Belonging

The inhabitants of Earth and Venus would not be able to exchange their living environments without the mutual destruction of both.
—Immanuel Kant, *Universal Natural History and Theory of the Heavens*

If there is life on Mars, we should do nothing with Mars. Mars then belongs to the Martians, even if the Martians are only microbes.
—Carl Sagan, *Cosmos*

On August 7, 1996, US scientists announced that they may have discovered life on Mars. The evidence: a four-pound chunk of meteorite found by a research team in the Allan Hills region of Antarctica and tagged "ALH84001." Determined to be of Martian origin due to traces of gases matching the Martian atmosphere, researchers later discovered trace amounts of polycyclic aromatic hydrocarbons (which often have biological origins), magnetite and iron sulfide (which are related to bacterial action), and carbonate globules (which are similar to substances produced by terrestrial bacteria). "Although there are alternative explanations for each of these phenomena taken individually," the research team concluded, "when considered collectively, we conclude that they are evidence for primitive life on early Mars."[1] As Richard Zare, one of the members of the discovery team, explained, the rock was not just evidence of life elsewhere—it shifted scientific thinking about life itself: "from life being special to life being ubiquitous."[2]

At the press conference announcing the discovery, as scientists and NASA administrators smiled and posed for pictures, ALH84001 sat before them, perched magnificently on a velvet pillow and encased in a glass box as though it were the Hope Diamond. Rocks do not talk, but this one spoke. Bill Clinton gushed in a speech on the White House lawn:

> Today, rock 84001 speaks to us across all those billions of years and millions of miles. It speaks of the possibility of life. If this discovery is confirmed, it will surely be one of the most stunning insights into our universe that science has ever uncovered. Its implications are as far-reaching and awe-inspiring as can be imagined. Even as it promises answers to some of our oldest questions, it poses still others even more fundamental. [The] fact that something of this magnitude is being explored is another vindication of America's space program and our continuing support for it, even in these tough financial times.[3]

The first bold claims about ALH84001 eventually weakened and then became the subject of controversy.[4] However, although the question of whether *indigenous* life has ever existed on Mars remains open, we can be fairly certain there has been life on Mars, attests NASA scientist and astronaut John Grunsfeld, "because we sent it there."[5]

Each time humans land (or crash) something on Mars, that something delivers a small number of microscopic hitchhikers to its surface. Just as the potential microbes in ALH84001 became a matter of cosmic significance when they were thought to come from another world, when placed in an extraterrestrial context, these tiny forms of terrestrial life raise questions of a different sort. What would it mean if terrestrial bacteria found an ecological niche on Mars and multiplied, overtaking potential native biota? Or what if we could establish that Mars is almost certainly *devoid* of life, past and present? Is it then our responsibility to keep it that way? Or, if life is a universal good, is it humanity's moral obligation as life's self-appointed ambassadors to spread it throughout the universe?

These questions are not just fodder for overactive imaginations. Policies prohibiting interplanetary contamination may be found in the earliest international treaties guiding space exploration, and space programs have spent hundreds of millions of dollars to prevent the scenarios I just described. Some of the most influential figures in space science, like Carl Sagan, have argued that if life were to be found on Mars, it should take precedence over any possible human interest. We should protect that life, he contends, by staying away. Others find this idea ridiculous. Robert Zubrin, president of the Mars Society and one of the most vocal advocates for the colonization of Mars, has called Sagan's position "immoral and insane."[6] But what does it mean to consider potential Martian microbial life worthy of moral consideration to the point that we would abandon one of humanity's boldest fantasies in order to protect it? Why would

Martian microbes be "a treasure beyond assessing,"[7] as Sagan described them, but the microbes on your hands may be washed down the drain without a second thought?

To untangle these questions, it is first necessary to look closely at the tiny life-form at their center. For many years, scientists classified all life on Earth into two primary domains: eukaryotes (which include plants, animals, fungi, and single-celled organisms like paramecia) and prokaryotes (which include bacteria and other microbes). In 1977, a team of researchers led by Carl Woese discovered a third domain: archaea, a category of microorganisms that include extremophiles living in areas originally thought to be inhospitable to life, such as those devoid of oxygen or with temperatures above the boiling point of water—the very organisms most likely to survive heat-sterilization procedures, the long trip through space, and the extremes of the Martian surface.[8]

The understanding of microbial life *as* life tends to have an implied asterisk, however. When people revere Earth's oldest life forms, for example, they tend to focus on animals like tortoises or plants like California's bristlecone pine—one such tree is over five thousand years old. In 2007, however, a team of researchers discovered still-living bacteria in Siberia that are over *half a million* years old, a discovery that warranted barely a mention in the news.[9] It's not that we don't understand microorganisms to be forms of life. It's that we forget about them. "Life," as Lynn Margulis once put it, usually refers to "big things like us."[10]

If it is difficult to consider microbial life *as* life, it is even harder to grant it moral considerability. Ever since Louis Pasteur gave germs a bad rap in the nineteenth century, microbial life has taken shape in medical practice and the popular imagination as a form of life that not only may be killed but *must* be killed to protect human life. Arguments to the contrary have often been put forward as ridiculous—the ultimate limit of ethical extensionism.[11] But if the popular attitude seems to be shifting, it is not because we recognize the inherent value of microbes or consider them to be fellow creatures, as we saw in deep ecology's rhetoric, but because they are understood as necessary symbionts for human life.[12] While some people may be waking up to the instrumental value of microorganisms for human health, in other words, the case for their intrinsic value is much harder to make.

The microbiologist Charles Cockell, one of the few scholars to argue for the intrinsic value of microorganisms, argues that the reason microbes are rarely considered morally significant is simply because they are so small. "Size bias," he writes, is "pervasive" in "environmental thinking and practical environmental policy," which may be one of the reasons that campaigns exist to protect "charismatic megafauna" like whales or polar bears rather than any species of microfauna, no matter how rare or beautiful.[13] Cockell suggests that microbes do not feature in environmental ethics or policy because of their physical size, but it's more than that. Because they are small, *they are not thought to matter*.

Arguments on behalf of microbes are first and foremost arguments of *mattering*; that is, they are appeals to what rhetoricians call "magnitude," the attempt to establish "the gravity, the enormity, the weightiness of [a subject,] a sense of significance that may be glimpsed and recognized by others," so they may "attend to it, to engage it, and to act upon it," writes Thomas Farrell. As such, magnitude is not just a trope or strategy, Farrell suggests, but a fundamental element of rhetoric, the "fine and useful art of making things matter."[14] By focusing attention and establishing significance, magnitude sets the ground for both ethics and politics, and it also adds to rhetoric, as Debra Hawhee explains, an aesthetic dimension, "in the ancient sense of sense perception."[15] Magnitude, in other words, makes things perceptible, and it does so by creating a sense of relation. As Farrell writes, "any matter of quality and importance must, in a rhetorical sense, be presented to us proportionally, on a human scale, so that it may be 'taken in.'"[16]

When we first encounter magnitude, Farrell writes, we don't immediately tend to questions of value: "All magnitude says is, 'Hey, look at this! This is important!' It is, as the phenomenologists might say, a call to attention, an exclamation point. All the problematical questions of value arise when we pause to say, 'Why?'"[17] Keeping in mind the philosopher Kelly Smith's point that most ethics discourse about Martian microbes takes place outside the discipline of ethics, this chapter examines the question of "Why?" from the perspective of a group of people we might call "astrobioethicists," which includes scientists as well as philosophers, and writers of popular science and science fiction who grapple with questions regarding the moral considerability of microbial extraterrestrial life.[18] Fiction may be an especially rich site of this grappling because it provides us, Susan Squier writes, with "a thick description that can provide a ground on which to debate moral questions."[19] Moreover, as we will see, fiction has played a direct role in shaping science, policy, and the public understanding of the significance of extraterrestrial life. In what follows, I first provide a brief history of the field of exobiology (now usually called "astrobiology") and show how the search for life elsewhere required rethinking the meaning of "life" at the center of the life sciences. I then look closely at discourse about two forms of interplanetary contamination—back contamination and forward contamination—that emerge as matters of concern for those who are committed to ethical space exploration.

Arguments about astrobioethics are grounded in magnitude using the many topoi that Farrell tells us that magnitude may take: unique/ubiquitous, near/far, infinite/infinitesimal, gigantic/tiny, somber/ridiculous. As we will see, these arguments establish a network of significance among microbial life, human life, and life itself that is given additional magnitude by place—that is, by a particular planetary context.[20] "Life on Earth" often functions as a synonym for "all life" or "life itself." But if life were to be found on other planets, that is, if life is discovered to be a universal rather than a terrestrial phenomenon, then "life on Earth" shifts to a deictic marker: life *on Earth*.[21] Of all the chapters in the book, this one drifts the farthest from the neighborhood we have been wandering in, so to speak. But by doing so, it also may bring us closest to the questions we have been orbiting all along: What is life? Why does it matter?

Life Everywhere: The Search for a Universal Biology

The discovery of ALH84001 provided vital public attention (and money) to NASA in the mid-1990s, but one of its most important effects may have been its work to establish extraterrestrial life as a valid object of scientific research. Until this point, writes Zubrin in an infantilizing analogy, "people interested in looking for life on Mars were viewed as like grad students who never grew up."[22] After Clinton's call for more research in his speaking-rock speech, NASA created the Astrobiology Institute in 1998 and appointed the Nobel laureate Baruch Blumberg as its first director. As the astrobiologist Jeffrey Bada recalls, extraterrestrial life seemed to burst from the margins of science to the mainstream at this time, becoming the "hottest topic around."[23] But although there was a rush of attention to extraterrestrial life at the turn of the twenty-first century, scientific interest in the topic was not new, nor was it a fringe interest. The search for extraterrestrial life has been part of mainstream space science and space policy from the very beginning.

Almost immediately after NASA formed in 1958, the agency began funding research into the question of whether there might be other forms of life in the universe and also began deliberating about the scientific, ethical, and political questions such life might provoke. This was due in large part to the efforts of

Joshua Lederberg, a molecular biologist who won the Nobel Prize in 1958 for his work on the transfer of genetic material between bacterial cells and who is known as one of the founders of the field of exobiology. "I was the only biologist at the time who seemed to take the idea of extraterrestrial exploration seriously," Lederberg remembers about this moment. "People were saying it would be a hundred years before we even got to the moon."[24] As a teenager, Lederberg first became entranced with the idea of extraterrestrial life after hearing Orson Welles's 1938 radio adaptation of *War of the Worlds*, which was set only sixty miles from his childhood home in Montclair, New Jersey. (He claimed not to have been fooled by the broadcast and thought the "subsequent news reports of public panic were part of the Halloween spoof.")[25]

While on a Fulbright in Melbourne twenty years later, Lederberg, like many others in the Southern Hemisphere, watched the launch of Sputnik with his own eyes. The event made a deep impression on the young scientist. Lederberg remembers that during a dinner with J. B. S. Haldane in Kolkata the following month, both men "shared the lament that this magnificent scientific opportunity, the beginning of human exploration of space, would likely be marred by the geopolitical competition, that it would be used for propaganda demonstration rather than scientific inquiry. Furthermore, we might have to take measures to protect the moon and other planets from inadvertent radioactive or biological contamination arising as byproducts of the circus."[26] When Lederberg returned to the United States later that year, he attended a party where the conversation turned to the Soviet Union's launch of dogs into space—"if one of these landed on the moon, it would contaminate the moon."[27] Shortly thereafter, Lederberg and his colleague Dean Cowie urged spacefaring nations to take the biological risks of space exploration seriously, in an article titled "Moondust" that they sent to both *Science* and the *New York Times*.[28]

Lederberg's initial worries about lunar contamination percolated into a broader interest: What would the discovery of extraterrestrial life mean for the study of biology? Humans have always wondered at the "demons which lurk beyond the Pillars of Hercules," he wrote in his first article on the subject, but merely as "adventuresome amusements" and "amateur delights." With the development of space exploration, the time had come for extraterrestrial life to become the subject of "fruitful exploration and dispassionate scientific analysis."[29] After shuffling through a few names (such as "cosmic microbiology" and "lunar biology"), Lederberg christened the new interdisciplinary science "exobiology."[30] In 1958, Lederberg persuaded the National Academies of Science to expand a small subpanel of the Space Sciences Board into two research groups: WESTEX, a union of scientists on the West Coast, and EASTEX, its eastern counterpart. In 1960, EASTEX and WESTEX merged into the "Committee 14 on Exobiology," with Joshua Lederberg serving as its first chair.

Exobiology offered something of incalculable value to the life sciences. Because "'life' until now has meant only terrestrial life," Lederberg wrote, efforts to establish theoretical biology as a field with the epistemological heft of theoretical physics or theoretical chemistry were doomed from the start; the life sciences have had to settle for the "rationalization of particular facts" rather than the search for universal principles. In a hierarchy of knowledge that prizes form and consistency—what the theoretical biologist Robert Rosen calls "syntactic" truth, irreducible and irrespective of particulars—the life sciences will always lag behind.[31] Unlike the constants of physics, which are believed to hold across the universe (or at least have been tested off planet) biological principles cannot be said to be universal, since the claim is undercut by one critical unknown: Are actions like reproduction, natural selection, and metabolism characteristics of all life or merely life as it has evolved in relation to Earth's specific environment?[32] Does all life have a carbon base or just life as it has emerged on this planet? Exobiology, then, would seek a definition of life that is abstract enough to apply to life on Earth *and* life everywhere else. "Only the perspective of comparative biology on a cosmic scale" would provide a field of vision large enough to ground such a definition, Lederberg believed.[33] To underscore this point, he even suggested that biology be split between exobiology and "esobiology"—the study of life on Earth—though the latter term never caught on.[34]

Lederberg's abiding concern with interplanetary contamination was inextricable from the development of exobiology as a science for a very simple reason. If you want to find and understand life elsewhere, it is imperative not to "find" the life you brought with you. Potential contaminants of science are manifold: personal (a desire for recognition), psychological (confirmation bias), material (a piece of dust on a lens), epistemological (a mistaken taxonomy), and ideological (religious or political bias).[35] The contamination of a celestial body like Mars with even the smallest amount of Earth life would thus add countless confounding variables to the search for life there, a point that Catharine Conley emphasizes in the interview that follows this chapter. Protecting the integrity of scientific research also serves another essential purpose: if we spent billions of dollars to discover life on Mars only to find life from Earth, public trust and support for space exploration would plummet.[36]

Interplanetary contamination not only threatened the search for life elsewhere but also the search for life else*when*: *How did life begin?*[37] Of particular interest to exobiologists was the theory of panspermia, an ancient idea attributed to the pre-Socratic philosopher Anaxagoras, who taught that life originated elsewhere and was "seeded" on Earth.[38] (Similar ideas are found elsewhere in the ancient world, such as Egypt, where carvings portray the creator god Osiris ejaculating across the night sky, and also in the Hindu *Rig Veda*).[39] At the turn of the twentieth century, however, panspermia shifted from an ancient curiosity to an active scientific interest, most notably through the efforts of Svante Arrhenius, who forwarded the idea in his best-selling book *Worlds in the Making*.[40]

In the "Exobiology" article, Lederberg proposed that research into panspermia was particularly at risk from irresponsible space exploration.[41] "In order to detect the 'life-bearing seeds,'" the historian Audra Wolfe explains, "scientists needed to be able to distinguish between man-made particles and unknown substances. The only way to ensure the detection of evidence of panspermia or other unfamiliar forms of life would be to keep the surfaces of celestial objects free from terrestrial contaminants."[42] Lederberg admitted that the plausibility of the panspermia hypothesis suffered from the assumption that microorganisms cannot withstand the vacuum, temperatures, and radiation of space long enough to travel from one celestial body to another. (This was 1960, more than a decade before the Woese team discovered archaea.) But while these factors made panspermia difficult as an experimentally useful hypothesis, Lederberg wrote, they still did not diminish its "immense significance for cosmic biology."[43]

Because of exobiology's vast scope, which covers the origins and evolution of life to its distribution in the universe, exobiologists were recruited from an array of disciplines, and their work, which required the cooperation of national space programs, made government support essential.[44] The new interdisciplinary field thus required collaboration not just among disciplines but among nations, and Lederberg was especially keen to establish channels of contact between American and Russian scientists. In a letter to Vladimir Timakov in 1958, Lederberg, speaking "only as a scientist and not as an official representative," sought to establish a line of conversation on interplanetary contamination, a concern that he argued ought to transcend professional and geopolitical rivalries. As space exploration ramped up after Sputnik, he argued, it was imperative to "rigorously exclude terrestrial contaminants from our spacecraft" until it could be determined conclusively that the risks of "detrimental effects . . . are small enough to warrant the relaxation of these controls." "The question of whether the planet Mars is inhabited by living organisms will be extremely pertinent to our conceptions of the origin and distribution of life in the universe," Lederberg wrote to Timakov, appealing to a shared scientific commitment that crossed national boundaries.[45]

Exobiologists found a receptive audience for their concerns. In 1958, the International Council of Scientific Unions (ICSU) recommended a set of standards regarding interplanetary contamination that was followed by a similar proposal by the National Academy of Sciences.[46] And perhaps most significantly, language about contamination was codified in the United Nations' 1967 Outer Space Treaty, which states that "parties to the Treaty shall pursue studies of outer space, including the Moon and other celestial bodies, and conduct exploration of them so as to avoid their harmful contamination and also adverse changes in the environment of the Earth resulting from the introduction of extraterrestrial matter and, where necessary, shall adopt appropriate measures for this purpose."[47] In response, NASA created the Office of Planetary Protection, the government body still tasked with guiding NASA policy and mission protocols regarding interplanetary contamination. International efforts against "biological interchange" are now the responsibility of the Planetary Protection Panel of the ICSU's Committee on Space Research (COSPAR), which evolves policy recommendations on contamination in line with the latest science.[48] These measures, as I will explain in the next section, were followed strictly during the Apollo 11 mission to the moon, but when the Viking 1 mission to Mars appeared to find no compelling evidence of life there, COSPAR relaxed its controls after concluding that stringent measures to reduce bioburdens on spacecraft, landers, probes, and rovers were based on "highly subjective speculation."[49] In their place, COSPAR recommended an approach to planetary protection that linked sterilization measures to mission type, which corresponds to a scale of permissible bioburden on spacecraft.[50] The strictest decontamination protocols are reserved for missions with the explicit purpose of life detection or those that may venture near areas with known or suspected water, such as the moons of Jupiter, which are thought to contain vast subsurface frozen oceans.

Contamination is an ethical issue by definition, as it names a kind of contact that ought not to take place, a violation of an a priori purity. Purity and contamination exist in relation, their borders erected by communities in concert with norms of order with which these terms are understood and given value.[51] Scientific contamination is about preserving the integrity of scientific knowledge and, in so doing, upholding the primary value system that makes science, well, *science*. In discourse about scientific contamination, Mars takes shape as a clean room, a technology to clarify scientific vision, a way of reducing noise so that life's signal, if it exists, can emerge. It is fundamentally an instrumentalist ethic in which celestial bodies and their potential inhabitants, however small, are valued for the kinds of knowledge claims they can support. In the case of exobiology, even a single microorganism has the potential for cosmic significance—depending, that is, on where it comes from.[52]

Life on Earth: Back Contamination

> No one would have believed in the last years of the nineteenth century that this world was being watched keenly and closely by intelligences greater than man's and yet as mortal as his own; that as men busied themselves about their affairs they were scrutinized and studied, perhaps almost as narrowly as a man with a microscope might scrutinize the transient creatures that swarm and multiply in a drop of water. With infinite complacency men went to and fro over this globe about their little affairs, serene in their assurance of their empire over matter. It is possible that the infusoria under the microscope do the same.

Thus begins H. G. Wells's science fiction classic *War of the Worlds*, which first appeared in serial form in *Pearson's* magazine (UK) and *Cosmopolitan* magazine (US) in 1897 and has been adapted in many other forms since, including the famous radio broadcast that captured Joshua Lederberg's imagination in 1938. Wells's story has been seen as a commentary on everything under the setting sun of the nineteenth century: Percival Lowell's visions of canals on a "dying" Mars, germ theory, British colonialism, the Spanish-American War, industrialism, and many others. Not surprisingly, the story is also frequently invoked as a warning about the perils of back contamination, a scenario in which extraterrestrial life threatens life on Earth.

As the Martians attack using large tripod vehicles equipped with heat rays, they quickly eradicate the Earthlings in their path, extracting their blood for food. As the tripods rampage across the landscape, they seed life in their deathly wake: a red Martian weed that rapidly spreads and chokes out native vegetation. The Martians are so relentless and efficient at their gory work that all hope for the Earthlings seems to be lost. Then, for no apparent reason, the tripods slow and grind to a stop. The famous twist in the story is that the Martians are felled unceremoniously not by humanity's biggest weapons but by nature's smallest creatures: microbes, "the humblest things that God, in His wisdom, has put upon this earth." Humans have evolved in a life-and-death dialectic with microbes across countless generations by "virtue of this natural selection of our land," Wells writes.[53] But on Mars, there are no bacteria. Possessing no immunity, the indominable Martians are conquered by something like the common cold.

Late in the story, as the traumatized narrator emerges from hiding and surveys the ruined landscape, he describes a strange feeling: "I felt as a rabbit might feel returning to his burrow and suddenly confronted by the work of a dozen busy [laborers] digging the foundations of a house. I felt the first inkling of a thing that presently grew quite clear in my mind, that oppressed me for many days, a sense of dethronement, a persuasion that I was no longer a master, *but an animal among the animals, under the Martian heel.*"[54] Here we see a kind of bio-identification that is tied not just to life but to *place*. The Earthlings in *War of the Worlds* are not just human. Although the Martians appear to have a taste for human blood, it was mixed with the blood of other "still living creatures."[55] Earth plants, too, fall prey to the red weed, the ultimate invasive species—which grew "with astonishing vigour and luxuriance." In the presence of a common enemy, even terrestrial microbes transform from the oppressors of animal Earthlings to their saviors, "microscopic allies" joined with their fellow living things through a coevolutionary history in a shared place.[56] As I noted earlier, "life on Earth" often is used as a synonym for life itself, with the understanding that "life on Earth" is *all life that exists*; however, when rhetorically positioned in relation to potential life elsewhere, "life on Earth" becomes particularized as life *on Earth*.

War of the Worlds offers a rich portrait of a threat to the planet's very biosphere, or what we might think of as a bioexistential threat. While this idea is similar to mass extinction in some ways, if considered from a much longer sense of time, mass extinction is a temporary event, as organisms have always emerged and evolved to populate Earth again. In these events, life itself is not destroyed, just (just!) vast numbers of living things. Bioexistential threat, in contrast, is a threat to the very possibility of life on the planet. The precise nature of that threat is an object of considerable debate, however. Some scientists use the conclusion of *War of the Worlds* to argue that since terrestrial organisms have evolved no defenses against extraterrestrial pathogens, *any* contact poses a bioexistential threat. Others believe that because terrestrial and extraterrestrial life have not evolved together, extraterrestrial microbes cannot take the form of pathogens, because they have no prior biological relationship with a terrestrial host. Still others believe that extraterrestrial life poses no threat at all because having evolved in a radically different environment, it would be unlikely to survive long enough on Earth to be dangerous. In the previous section, we saw how "life on Earth" shifts from a universal to a particular signifier when considered in the context of the literal universe. In what follows, we see how bioexistential threat joins life in biological community across species, emplacing and joining living things as "biotic citizens" of the Earth itself.[57]

Perhaps the most spectacular enactment of back contamination fears were the planetary quarantine measures enforced during and after the Apollo 11 mission to the moon.[58] As the Apollo program took shape in the 1960s, NASA organized several meetings and committees on the risks that the mission might pose to Earth. The first conference on the topic featured representatives from many federal agencies, including the Public Health Service, Army, National Institutes of Health, US Department of Agriculture, and the Fish and Wildlife Service. In 1964, a new committee emerged called the Interagency Committee on Back Contamination (ICBC), which was tasked with conducting research on potential risks and creating policies and procedures of planetary protection for Apollo 11. The ICBC's efforts culminated in a 571-page report called the "Baylor protocol," which creatively imagined how the lunar mission might threaten Earth, such as the mutation of terrestrial microbes on the Apollo Lunar Module into potential pathogens or the return of moon plants capable, like Wells's red weed, of overtaking terrestrial ecosystems.

The Baylor protocol produced a stringent set of mission procedures for the Apollo astronauts. After leaving the moon's surface, Neil Armstrong and Buzz Aldrin left their dusty boots and gauntlets in the module and vacuumed each other before rejoining Michael Collins in the command module. There they removed their suits and sealed them in plastic bags. After the crew splashed down in the Pacific Ocean, a recovery team quickly passed the three men biological isolation garments, which they scrambled into while still bobbing in the ocean. The garmented astronauts were then sponged with bleach and transferred to a mobile quarantine facility (a converted Airstream trailer) on the USS *Hornet* (fig. 3), which was airlifted first to Pearl Harbor and then to the Lunar Receiving Facility at the Johnson Space Center in Houston. There the astronauts lived with a colony of white mice for three weeks until NASA determined

Fig. 3 | President Nixon welcomes the Apollo 9 astronauts in quarantine, July 24, 1969. NASA.

that neither the men nor their samples were infectious. Collins later commented that all of the precautions may have been worthless after the moment of splashdown: "The command module lands in the Pacific Ocean and what do they do? They open the hatch. You gotta open the hatch. All the damn germs come out!"[59]

Collins had company in his criticism. Many leading scientists at the time thought the extensive, expensive lunar quarantine measures were "hogwash," "preposterous," "absurd," and "ludicrous" because the risk of back contamination was nearly nonexistent.[60] The University of Chicago chemist Edward Anders even volunteered to swallow a sample of moon dust to make the point.[61] Joshua Lederberg, who privately thought the quarantine "ridiculous," publicly argued that it served two important functions: to protect the lunar samples from *terrestrial* contamination and to serve as a dress rehearsal for human exploration of Mars, which he believed much more likely to harbor life than the moon.[62]

Whatever small measure of actual protection the lunar quarantine served, it appears to have served primarily as a piece of safety theater for the US public.[63] While many people now remember the Apollo 11 mission as one of humanity's greatest achievements, nearly half of Americans did not support it at the time.[64] Their concern echoed arguments from many scientists and politicians (such as Dwight D. Eisenhower) that a lunar mission was a project of national vanity simply not worth the risk or cost, as well as from cultural critics, like the poet Gil Scott-Heron, who pointedly argued that the money spent to get "Whitey on the Moon" was desperately needed to address racial and economic inequality at home.[65] For critics, it was difficult to justify such a massive expenditure of resources for what was widely seen as a geopolitical performance of techno-military might. How much more difficult the case for Apollo became, then, when people began worrying that the mission posed a bioexistential threat.

These fears came from somewhere. One likely origin is a wildly popular work of fiction published just two months before Armstrong and Aldrin set foot on the moon: Michael Crichton's first novel, *The Andromeda Strain*, which chronicles the spread of a virulent extraterrestrial pathogen.[66] The book begins with two US military personnel searching for a crashed military satellite in Arizona. As they scour the area, the soldiers notice carrion birds circling the small town of Piedmont and drive there to investigate. When the soldiers are discovered dead—along with nearly all of Piedmont's residents—the military initiates Wildfire, an emergency protocol designed to mitigate biological threats of extraterrestrial origin.

Crichton lavishes detail on the extensive measures of sterilization in the Wildfire protocols. The process unfolds over several chapters, as the main character (a scientist so obviously modeled on Joshua Lederberg that he complained)[67] literally descends through five levels of intensifying decontamination at the Wildfire laboratory: immunizations and full body scans, immersion baths, irradiation and ultraviolet light, antibiotic suppositories, and even a procedure in which the outer layer of his skin is burned off. In case the barriers were breached, the final sterilization measure is the automatic detonation of a nuclear bomb. At the climax of the novel, the Wildfire team discovers that Andromeda has mutated and is eating away at the rubber seals of the laboratory, threatening to escape. Andromeda thus posed a bioexistential threat of the highest order: its new form endangered not only human and nonhuman animals (all laboratory animals subjected to it died) but even a nonliving, yet organic, material like rubber. When the nuclear safeguard is triggered, the Lederberg character races to shut it down after he realizes that the explosion could "spread the organism all over the surface [of the Earth]. There will be a thousand mutations, each killing in a different way."[68] The scientists prevail, and in an ironic twist, they discover that the Andromeda strain was not extraterrestrial in origin. It was a terrestrial microbe that had hitched a ride to space on a satellite and mutated while orbiting the Earth (which, if you remember, was one of the specific concerns outlined in the Baylor protocol).

The Andromeda Strain reads less like the what-if scenarios of science fiction and more like the what-happens scenarios of popular science. This was intentional. After Crichton sent the initial draft to Knopf, Crichton's editor told him to rewrite the manuscript so that it was "absolutely convincing" and read like a "*New Yorker* profile" rather than fiction. The effect of the change was that "lots of people thought it was true," Crichton later recalled.[69] It's not hard to see why: in the opening acknowledgments, Crichton describes the story as a "history of a major American scientific crisis," made possible by the release of "declassified" information and "interviews" with dozens of participants, who are listed by (fictional) name. As Crichton explains in the acknowledgments, his objective with *Andromeda* was to raise public awareness of how scientific crises arise and are handled, because "in the near future, we can expect more crises on the pattern of Andromeda."[70] The *New York Times* was not shy in questioning *Andromeda*'s literary merit, but it nonetheless commended Crichton's technical "craft" and predicted it would be a "sure bestseller."[71] It was. The book enjoyed a fourth printing a mere month after its June 8 release, and it was the featured selection of that month's Book of the Month club. A movie deal followed shortly thereafter.

According to the microbiologist Abigail Sayers, *Andromeda* created "a fear in the public mind that extraterrestrial specimens might introduce a new and unstoppable plague to the Earth."[72] While the lunar quarantine measures were in place well before the summer of 1969, there is no question that *Andromeda* incited public concern that the lunar mission posed a grave threat to humans and other life on the planet. A *New York Times* article on the Apollo 11 back contamination measures, for example, begins with an explicit reference to *Andromeda* and suggests that the "chilling account" found in Crichton's novel "will hardly gladden the hearts of the National Aeronautics and Space Administration for it dramatizes the dangers of 'back-contamination' that have suddenly become a subject of sharp debate on the eve of the Apollo 11 mission of the moon."[73] In the lead-up to the mission, NASA received thousands of letters from Americans terrified of moon germs. Even fifty years after *Andromeda*'s publication, it remains a "touchstone for planetary protection specialists in discussing matters of back contamination and the dangers that they present."[74]

Priscilla Wald suggests that *Andromeda* is one of a handful of "outbreak narratives" that illustrate the "cataclysmic consequences" of contagion on a planetary scale.[75] Outbreak narratives, like the contagious diseases they chronicle, are more than just stories of sickness, risk, and danger. They also create a sense of belonging, both through the production of a threatened population—a biologically imagined community—and through the mitigation of that threat, as microbes are figured as interspecies enemies: human-Us versus microbial-Them.[76] Though microbes and viruses famously do not respect the boundaries of nations, the "effort to contain the spread of a disease . . . is cast in distinctly national terms," Wald writes, most obviously through the practice of quarantine, which erects boundaries around the collective biological body.[77] While the image of the quarantined Apollo astronauts, like the mission itself, is unmistakably American, the lunar quarantine sought to protect more than just American life, more than even human life: back contamination was a threat, the *New York Times* explained, to all "life on earth."[78] Indeed, NASA's Office of Planetary Protection, the body tasked with developing and implementing back contamination procedures for US space exploration, names its mission as protecting not humans, specifically, but Earth's biosphere.[79]

As of this writing, the most likely mission with the potential for back contamination is NASA's plan to return samples from Mars, though it is still in the early stages of development. While the prospect of recovering extraterrestrial life on this mission is small, and the chance of encountering pathogenic extraterrestrial life even smaller, the probability of back contamination is still "nonzero," explains former Planetary Protection Officer John Rummel, "and the potential for such an entity to cause damage to the Earth's biosphere cannot be discounted."[80] Some people argue that the "nonzero" chance is simply not worth the risk. A small eclectic group called the International Committee Against Mars Sample Return (ICAMSR), for example, exists for the sole purpose of lobbying US and international scientists and policy makers that all Martian material should be studied in situ.[81] The group cites no less a figure than Carl Woese to justify their concern: "When the entire biosphere hangs in the balance," Woese warns, "it is adventuristic to the extreme to bring Martian life here. Sure, there is a chance it would do no harm; but that is not the point. Unless you can rule out the chance that it might do harm, you should not embark on such a course."[82]

Arguments in favor of stringent precautionary measures often use the example of invasive species to bolster their case. While *War of the Worlds* is frequently raised as a lesson to this end, others use examples from terrestrial human history. "The concepts involved in planetary protection are not unfamiliar to anyone who has studied the history of human exploration," Rummel argues, "whether through episodes like the introduction of the rat to Hawaii by the Polynesians, the more recent spread of the zebra mussel into the North American Great Lakes by bilge water from ships returning from Europe, or the more-widespread exchange of microbes by seagoing vessels."[83] "People didn't realize that kudzu would take over the entire [US] South," comments the microbiologist Benjamin Wolf. "We need to understand that sometimes scientists don't know things."[84]

Still others use the genocide of Indigenous peoples to illustrate the existential threat that back contamination poses to life on Earth, a point echoed by Catharine Conley in our interview.[85] While a staggering number of Indigenous people were directly murdered at the hands of European settler-colonists, far more were killed by the microbial allies on those hands—sometimes deployed intentionally—as diseases like smallpox swept across the Americas, unrestrained by the immunities that had been built over generations across the Atlantic. If this sounds similar to the plot of *War of the Worlds*, it is no accident. Early in the story, Wells makes the colonial theme of the story explicit, inviting his British readers to compare themselves to the peoples their nation has tried to eliminate: "Before we judge [the Martians] too harshly we must remember what ruthless and utter destruction our own species has wrought, not only upon animals, such as the vanished bison and the dodo, but upon its own inferior races. The Tasmanians, in spite of their human likeness, were entirely swept out of existence in a war of extermination waged by European immigrants in the space of fifty years. Are we such apostles of mercy as to complain if the Martians warred in the same spirit?"[86] This famous passage has been much argued over—despite the racist language that portrays the Palawa, the Indigenous residents of Tasmania, as an "inferior race," subhuman, and extinct, is it somehow a statement *against* colonialism? Or is it a sympathetic defense of colonialism when at the hands of "superior" beings? One thing is clear: in *War of the Worlds*, Wells wished for his British readers to reflect on their relationship to the rest of the world, both as citizens of a violent, imperial nation and as the self-appointed "masters" of all life on Earth. As the colonizers in *War of the Worlds* are knocked from their exceptionalist perch and squeezed under the Martian heel—an animal among animals—they become the colonized, and the predators, prey.

A bioexistential threat joins disparate forms of life in biological community, but it also emplaces that community. In response to the threat of invasion by the ultimate Them, a new Us emerges—life on Earth is unified *as life* but also as *life in a particular place*, autochthonous, life *from* the Earth. At the end of *War of the Worlds*, Wells includes a point of reflection to this end by once again asking his reader to imagine themselves in the Martians' place. "If the Martians can reach Venus," he speculates, "there is no reason to suppose that the thing is impossible for men, and when the slow cooling of the sun makes this earth uninhabitable, as at last it must do, it may be that the thread of life that has begun here will have streamed out and caught our sister planet within its toils. Should we conquer?"[87] The Martians' attempted conquest of Earth may have arisen out of necessity, a scenario of scarce resources in which the Martians weighed their own lives against the lives of Earthlings and saw fit to sacrifice us to ensure their survival. One day our planetary home also will be unlivable—perhaps while the sun is still shining—and, if humans still exist and have the means to do so, we may be forced to seek a new one. On this day, we will bring many other Earthlings with us: animals, plants, and, invariably, countless microbes. "Should we conquer?" Wells asks, which suggests that the answer may be *no*, even when the very survival of "the thread of life" is on the line. Perhaps the Martians—kin related by the solar system, as "sister planet" implies—might have lives worthy of respect. If Earth belongs to us, then perhaps Mars belongs to them, too.

Life on Mars: Forward Contamination

Joshua Lederberg recalls that his efforts were motivated by a "heartfelt concern for protecting outer space."[88] The article in which he laid out the case for exobiology, for instance, concludes with a section titled "Conservation of Natural Resources," which hints at a risk of irresponsible space exploration beyond the contamination of science. "History shows how the exploitation of newly found resources has enriched human experience," he writes, but "equally often we have seen great waste and needless misery follow from the thoughtless spread of disease and other ecological disturbances."[89] While "resources" suggests that celestial bodies ought to be protected for what they offer to humans, or what is known in environmental discourse as a "conservationist" approach, the term "misery" and Lederberg's worry about ecosystems implies that there may be a reason to protect Martian life for its own sake, a sentiment that edges close to preservationism, the belief that natural areas and wildlife ought to be withheld from development and protected from violation because they have inherent value.[90]

The scenarios of forward contamination that I examine in this section—which extend from simply bringing an inadvertent microbe to Mars to large-scale human settlement and the terraforming of its surface—contend directly with the question of preservationism and its implications for the future of space exploration. In forward contamination, the protagonists and antagonists of *War of the Worlds* are reversed: the Martians are not the threat but the threatened. In these arguments, which often rely on comparisons to imperial violence and environmental destruction on Earth, we see collisions between a number of ethical approaches, each of which places a different moral patient at the center.[91] Thinking of Mars as valuable because it is an object of scientific knowledge or the source of natural resources or as land for future human settlement reflects an anthropocentric approach as well as an unmistakably Western and colonial mindset.[92] As we saw in chapter 2, biocentric and ecocentric frameworks, in contrast, extend moral consideration beyond humans, but in so doing, they raise questions about the reach of that consideration. In this section, I first examine how these ethical frameworks are deployed in arguments that assume that Mars has the potential to harbor microbial life. In the next, I turn to arguments that assume that Mars is devoid of life, where we find a fascinating, unstated premise that flows through the entirety of this book, a claim so obvious (and, thus, so powerful) that it usually requires no argument: *life is good.*

In the best-selling book *Cosmos,* Carl Sagan offers an explicit preservationist argument about Mars—a proposal that, if developed into policy, would have clear implications for the exploration of other celestial bodies. "If there is life on Mars," Sagan argues, "we should do nothing with Mars. Mars then belongs to the Martians, even if the Martians are only microbes. The existence of an independent biology on a nearby planet is a treasure beyond assessing, and the preservation of that life must, I think, supersede any other possible use of Mars."[93] This conditional argument features several intriguing premises. The first positions Martian microbial life as not just intrinsically valuable but *invaluable,* such that no human interest in Mars can be weighed against it. Sagan's use of the term "independent biology" is also notable—he assumes that life on Mars represents what the NASA astrobiologist Christopher McKay calls a "second genesis"—a line of life that has emerged independently from life on Earth.[94]

From an ethical point of view, argues McKay with Margaret Race and Richard Randolph, the cosmic significance of a second genesis demands a preservationist approach: "the need to preserve a life-form, however lowly, must be more compelling if that life form represents a unique life-form with an evolutionary history and origin distinct from all other manifestations of life."[95] Mark Lupisella agrees, calling potential life from a second genesis "a treasure beyond measure—a unique jewel of the rarest sort," which makes its intentional or unintentional extinction a "crime of cosmic proportion."[96] Here we see moral consideration tied explicitly not necessarily to an organism, per se, but to its place and manner of origin. The same Martian bacterium, in other words, is not only variably significant but *a different moral patient* if it (a) originated on Earth and was the product of recent contamination, (b) originated on Earth long ago, was ejected into space from collision impact, and traveled to Mars, (c) has the same origin as terrestrial life, whether by panspermia or the hand of God, or (d) emerged and evolved independently. Yet Sagan's brief argument extends beyond value. It also provocatively raises the issue of territoriality between indigenous Martian life and Martian land. This is the case "even" if the Martians are not intelligent or even sentient beings but "only" microbes, Sagan suggests.

Using these biocentric approaches as a foundation, Randolph and McKay propose a guiding astrobioethical principle for space exploration that grants "intrinsic value to all living organisms," terrestrial and extraterrestrial alike.[97] Such a universal principle is necessary, they write, because moral reasoning is culturally and historically contingent. They thus propose life itself as a "common anchor-point," "that is, a value that is common to this multitude of ethical perspectives around the world": "We believe that this common anchor-point is actually very simple and straightforward—it is simply, life. Despite multiple differences in culture, religion, and custom, living is one reality that all persons share in common with one another as well as with non-human animals, plants and microbes here on Earth. If we discover extraterrestrial organisms, whether it is a humble microbe on Mars or a vastly superior space traveler from another planet, *we will share life in common*."[98] Using bioidentification to establish a common substance, employing a biocentric ethics that grants moral consideration to all forms of life, applying the precautionary principle, and following what they call the "cosmic golden rule"—treating "inferior" extraterrestrial life in such a way that we would like to be treated by "superior" extraterrestrial life, an echo, once again, of *War of the Worlds*—Randolph and McKay hope for a universal ethical principle to guide right relations among the community of life in the universe.

The value of an extraterrestrial microbe is located in the life it holds in common with humans, as Randolph and McKay state explicitly. However, much like we saw in the previous section, even the humblest Martian microbe is not just *any* microbe. The magnitude of moral considerability in this argument is also tied to its difference: in this case, emplacement on a celestial body other than Earth. Bioidentification here works as a kind of synecdoche whose rhetorical force is charged with magnitude: the humble microbe on Mars, unlike the humble microbe on Earth, comes to stand in for life itself, and its significance expands accordingly.

Life in the Dust

But what if Mars has no life? What if it never had life? Is the planet then devoid of intrinsic value and thus not worthy of consideration or protection?[99] The claim that life itself is sufficient for moral standing is controversial enough. Even Paul Taylor did not extend moral consideration to nonliving things in *Respect for Nature*; sand, he argued, cannot have a good of its own.[100] Carl Sagan's argument ceding Mars to Martians is rooted in Martians—even microbial Martians—as *living* things. Martians, in other words, cannot be rocks. Perhaps Sagan would agree with the astrobiologist David Grinspoon, who answers his essay title "Is Mars Ours?" in the affirmative: "Mars does not belong to 'America,' nor to Earth, nor to human beings," Grinspoon writes. "But if by 'we,' we mean 'life,' then yes, Mars belongs to us because this universe belongs to life."[101] "Life has precedence over non-life," writes McKay. "Life has value. A planet Mars with a natural global-scale biota has value *vis-à-vis* a planet with only sparse life or none at all."[102] Zubrin agrees. "In securing the Red Planet on behalf of life," he writes, "humans will perform an act of improving creation so dramatic that it will affirm the value of the human race, and every member of it. There could be no activity more ethical."[103] For each of these thinkers, life appears as the unstated warrant that authorizes Martian exploration and settlement as not just a human right but a "responsibility as members of the community of life itself," as Zubrin puts it.[104]

As we saw in chapter 1, the biotic/abiotic dichotomy that characterizes the Western ontological economy is deeply rooted in the value we place on life. But, as Erin Daly and Robert Frodeman argue, in an echo of that chapter, "abiotic nature can also have value through the *relatedness* of nature and natural objects to human beings. We may be confident that rocks do not think or have values of their own. But humans can nonetheless value rocks for their own sake—they can be experienced as beautiful, sublime, or sacred." Alan Marshall, for example, insists that "nature is not static," even in abiotic worlds: "Myriads of dynamic physical, chemical, and geological phenomena permeate lifeless planets. The turbulent atmosphere of Neptune, the volcanic activity on Jupiter's moons, and the chemical reactions of the surface-atmosphere interface of Venus could fulfil many definitions of what it is to be 'alive.' . . . Although [Mars] might seem to be a great useless hunk of red rock to us, humans could, in the view of Martian rocks, be merely living organisms who are yet to attain the blissful state of satori only afforded to non-living entities."[105] Because the "exploration of the universe is no experimental science, contained and controlled in a laboratory, but takes place in a vast and dynamic network of interconnected, interdependent realities," Daly and Friedman argue, it is crucial that scientists and policy makers interested in these questions engage in regular conversation with humanists and others who regularly grapple with questions of meaning and value.[106] At this point, I know of no space scientist or policy maker who proposes extending moral consideration to the blissful Martian regolith. But as is often the case when we wonder about worlds beyond our own, guidance might be found in science fiction.

Green Mars, the second volume of Kim Stanley Robinson's magisterial Mars trilogy, opens with a lesson on the meaning of life in the universe. Walking on a beach, Hiroko Ai, one of the original hundred settlers of Mars, picks up a seashell and invites her Martian-born human pupils to examine its intricate pattern:

> The dappled whorl, curving inward to infinity. That's the shape of the universe itself. There's a constant pressure, pushing toward pattern. A tendency in matter to evolve into ever more complex forms. It's a kind of pattern gravity, a holy greening power we call *viriditas*, and it is the driving force in the cosmos. Life, you see . . . And because we are alive, the universe must be said to be alive. We are its consciousness as well as our own. We

> rise out of the cosmos and we see its mesh of patterns, and it strikes us as beautiful. And that feeling is the most important thing in all the universe—its culmination, like the color of the flower at first bloom on a wet morning. It's a holy feeling, and our task in this world is to do everything we can to foster it. And one way to do that is to spread life everywhere.[107]

Life is central to Hiroko's worldview, a religious devotion called the "areophany." Humans in the areophany are life's ambassadors and missionaries—they are bound by citizenship in a living universe and a responsibility to spread the "holy greening power" that is its "driving force." Thus, it is on life's behalf that Hiroko and her fellow areophanites join with Martian "Greens" (those in favor of terraforming Mars) against the "Reds" (those in favor of preserving Mars as it is).

The fight between the Red and Green worldviews is anchored in a set of trilogy-spanning debates between the original Green, the biologist Sax Russell, and the original Red, the geologist Ann Clayborne. Ann is adamant that Mars is its "own place," a thing of intrinsic value. "We are not lords of the universe," she insists. "We're one small part of it. We may be its consciousness, but being the consciousness of the universe does not mean turning it all into a mirror image of us. It means rather *fitting into it as it is*, and worshiping it with our attention."[108] Sax, in contrast, declares Mars "dead"—merely a "speck of random matter" in the universe.[109] If Mars is beautiful, Sax insists, its beauty is found in the imaginations of science fiction writers and the data of scientists, echoing Daly and Frodeman's point earlier, not in the "basalt and oxides" of the Martian surface. For Sax, Mars is a "blank red slate" waiting for the hand of human innovation to lift the planet to its full potential.[110] While not one of Hiroko's devotees, Sax nonetheless believes that Mars is significant only in relation to human perception and valuable only with the addition of life, "the most beautiful system of all."[111] Ann, in contrast, "believed in some kind of intrinsic worth for the mineral reality of Mars; it was a version of what people called the land ethic, but without the land's biota. A rock ethic, one might say. Ecology without life."[112]

These two positions, Red and Green, are fundamentally at odds—and the tension eventually breaks into something of a civil war. But in the areophany, we see a different way of thinking that joins the life ethic and the rock ethic as inseparable, intertwined perspectives. This idea is vividly illustrated in Hiroko's areophany ceremony described in detail in *Red Mars*. As her followers gather in a circle, Hiroko gives them each a handful of Martian regolith. "This is our body," she says and begins to eat the red dust in her hand. Though revering life as a holy obligation, Hiroko's areophany is centered not on life but on the abiotic/biotic interdependence of viriditas and *kami*, a Shinto term that refers to "the spiritual energy or power that rested in the land itself":

> *Kami* was manifested most obviously in certain extraordinary objects in the landscape—stone pillars, isolated ejecta, sheer cliffs, oddly smoothed crater interiors, the broad circular peaks of the great volcanoes. These intensified expressions of Mars's *kami* had a Terran analogue within the colonists themselves, the power that Hiroko called *viriditas*, that greening fructiparous power within, which knows that the wild world itself is holy. *Kami, viriditas*; it was the combination of these sacred powers that would allow humans to exist here in a meaningful way. . . . When Michel heard Evgenia whisper the word "combination" . . . all antinomies collapsed to a single, beautiful rose, the heart of the areophany, *kami* suffused with *viriditas*, both fully red and fully green at one and the same time.[113]

Kami speaks of the value of the abiotic elements of Mars not because of their usefulness, their beauty, or their size. Though found "most obviously" in spectacular, extraordinary elements of the landscape, *kami* is also manifest in the ordinary dirt in the settlers' hands. "This is our body," Hiroko tells her followers, reminding them—and us—of the a priori consubstantiality of biotic and abiotic elements that requires no ritual for transubstantiation. Identifying not as Terran but as Martian, making new life and lives there, opening their bodies to its red materiality and the new worlds it makes possible, Hiroko's followers attest that Mars does not belong to them nor even to life. They belong to Mars.[114]

After the announcement of possible life in the Allan Hills meteorite, newspapers breathlessly wondered if at long last there was an answer to one of our oldest questions: Are we alone in the universe? To imagine other worlds on other planets is motivated by a desire for communion; to contemplate Earth's place within the solar system and universe is a salve for cosmic loneliness. "To speak of a 'planet' is never to speak of an isolated body," writes Lisa Messeri. "We will only be less alone if we can connect with, imagine *being* on, and thus being in place on, these other worlds."[115] This is all the more true when it comes to imagining life there—and being *with* it, even if that life is microscopic and millions of miles away. A microbe can be nothing or everything, not-life or life itself, ancient kin or a mere germ to be washed down the drain. To be *with* such life depends on the many ways that "I" and "we" come to be, the substances we believe we share, and the thrums of resonance and dissonance that shape the ethical relations between us.[116]

Coda: Looking Across Worlds

One hundred years before NASA's triumphant announcement about ALH84001, the American polymath Percival Lowell claimed that he had found proof of life on Mars. His evidence: an intricate series of what looked to be canals crisscrossing its surface. The formations were too straight to be natural, Lowell argued, which meant that they must have been made by someone.[117] But while the scientific community remained skeptical of Lowell's canals and their putative builders, Lowell remained firm in his conviction for many years, having seen the structures with his own eyes. Using a giant refracting telescope at his observatory in Flagstaff, Arizona, Lowell also claimed to have seen a pattern of formations on Venus, which he described as "spokes" around a central hub (fig. 4). In fact, Lowell claimed to have seen patterns on not just Mars and Venus but on nearly every celestial body he observed. In 2003, a paper published in the *Journal of the History of Astronomy* speculated that the way that Lowell had calibrated his telescope may have effectively turned it into a giant ophthalmoscope, which you might know as the blindingly bright tool an eye doctor uses to examine your optic nerve. Lowell looked into the heavens, and it seems that what he saw there was the back of his own eyes.[118]

In the introduction of this book, I discussed the power of the Overview Effect, the sublime feelings of interconnection produced by viewing Earth from above. Does gazing up at Mars evoke something similar? I've asked myself this question many times as I pore over gorgeous high-resolution images of the Red Planet: the long slice of the Valles Marineris, the longest valley in the solar system, named not after a human but a machine, the Mariner 9 orbiter that conclusively proved Lowell's canal theory to be wrong; the ice capping each pole, a swirling mixture of frozen water and carbon dioxide; Olympus Mons, the gargantuan shield volcano that is three times as tall as Mount Everest; the rusty regolith, in which I try to contemplate the *kami* that Hiroko exalted as well as the "blissful satori" of the rocks scattered across the silent surface. I have tried to lose myself in these images of Mars in the way that Joshua DiCaglio describes losing himself in an image of Earth.[119] But what I feel is . . . nothing.

Fig. 4 | Percival Lowell observes Venus at his telescope in Flagstaff, Arizona. 1914. Public domain, image provided by Wikipedia.

The psychiatrist Nick Kanas and the psychologist Dietrich Manzey speculate that interplanetary travel may produce something of the *opposite* of the Overview Effect, something they term the "Earth-out-of-view phenomenon." They suggest that the act of leaving Earth behind—the point where the vast blue marble shrinks to nothing but the "pale blue dot" that enraptured Carl Sagan—may produce feelings of isolation and loneliness so intense that depression, psychosis, and suicidal thinking may result.[120] Like many other ideas in this chapter, this one has been thoroughly explored in science fiction. *Red Mars*, for example, features the character Michel Duval, a French psychiatrist who begins to lose his mind. Tormented with regret over his decision to leave Earth, Duval is consumed with thoughts of his home in Provence, "a living landscape, a landscape infinitely more beautiful and humane than the stony waste of this reality."[121] Duval misses Earth so much that he began to dissociate, thinking of himself and his fellow colonists as unfeeling computer programs, until the orgiastic conclusion of Hiroko's areophany gives him rebirth. If there is something like a Mars effect, it might be a feeling of alienation—perhaps the definitive example of it. Alienation, after all, is the separation of things that properly belong together.

Yet in the process of writing this chapter, thinking incessantly about Mars and the passionate hunt for even the tiniest speck of life there, and what it might mean, I have found myself looking not up but down—down, drawn to life in places I had never noticed it before, or at least hadn't *apprehended* as life before, as Butler would have it: purslane insisting on itself in the cracks of sidewalks, tiny sprouts of oaks in damp gutters, the living surfaces of lichens and molds and yeasts, the great horned owls conversing through the treetops in the early hours of the morning, the sudden mushrooms in my front yard driven by the unseen force the Potawatomi people call *puhpowee*. I have noticed life, and I have *marveled* at it, marveled at just how much life there is *everywhere* on this warm, blue and green planet.[122] There is life inside us and on us, life moving and growing and breathing and dying all around us, life in the high atmosphere, in the boiling vents in the ocean, and miles down into the rock beneath our feet. *There is life everywhere*. Looking up at Mars, I realize that, like Lowell, I have been seeing something else, or at least something familiar in a new way.[123] Maybe we are always looking down on Earth, even when we are looking up. The question, then, is what we learn to see there.

Signs of Life | A Conversation with Catharine Conley

Dr. Catharine Conley received her PhD in plant biology from Cornell University in 1994 and obtained a postdoctoral position at the Scripps Research Institute as a National Institutes of Health fellow studying proteins involved in muscle contraction. In 1999, she became a research scientist with the NASA Ames Research Center studying unloading muscle atrophy. Her later research focused on the evolution of motility, particularly animal muscle. One of her experiments was on board during the Space Shuttle *Columbia* disaster. The experiment, the fourteenth Biological Research in Canisters (BRIC-14), survived reentry, and the nematode cultures were found to be still alive. This experiment enabled a research program in C. *elegans* spaceflight biology that continued for two decades. In 2006, Conley was appointed NASA's Planetary Protection Officer (PPO) at NASA Headquarters, a position she held until 2017.

JJ: Could you begin by talking a bit about your background?

CC: Yes, I did my undergraduate degree at MIT. In my senior laboratory research class, one of the other students asked, "So . . . we're working with *Neurospora crassa*. Is that a plant or an animal?" As if those were the only two options! The whole time at MIT, nobody had learned any taxonomy, nothing about the history of biology on Earth. As a kid, my dad used to read me animal encyclopedias, so I was aware of animal taxonomy, but plants were something I didn't know about. I decided to take some classes about them but had to do it at Harvard, since MIT just didn't teach taxonomy. Plants are amazingly durable and more capable than animals. Animals can just run away from things. Plants cannot run away. Fungi cannot run away. So they have actually quite a few more biological capabilities than animals do. Learning that plants have so many capabilities is what led me to do plant biology for my graduate work. Also, plants don't mind if you cut parts off of them. It's called "pruning." You don't have to kill them to get your scientific material. [*Laughs*]

JJ: So, how did you find your way into astrobiology?

CC: My father, a mathematician, was working in dynamical systems. When he was a student with Jürgen Moser, he was working on dynamical systems related to orbital mechanics. He was consulting with NASA to solve the special case of the three-body problem that had been calculated for the Apollo program. My dad's mathematics were able to solve the special case rigorously: a theoretical solution, an actual mathematical solution, rather than a calculated solution. After he died, his advisor got the Wolf Prize, the mathematical equivalent of the Nobel Prize, for that work. And my dad's math is still used in orbital mechanics. So the whole time I was growing up, I was aware of working for NASA. My dad would have my brother and me run around a tree, imitating the moons of Mars, so he could try to see how tidal locking worked, and things like that. So I have a feeling for orbital mechanics from a somewhat practical standpoint.

When I was working on plant biology as a graduate student, I started to study actin cytoskeleton. For my postdoc, I wanted to work on actin cytoskeleton: I went looking for a biological function, a protein function, that would be needed in plants that people understood in animals, because plant biology doesn't get nearly so much funding as human biology. A lot of things get discovered in animals and then converted over to plants. The lab I was in as a graduate student actually was an example of the opposite, because that lab discovered RNA editing in plants, in petunias. Maureen Hansen was one of my two major advisors: her lab actually originally discovered RNA editing. And then, it was recognized in animals *after* plants. That's one of the very rare cases where something was discovered in plants and then transferred to animals.

As a postdoc, I saw the NASA Pathfinder website. I just emailed a contact on the website to say, "Okay, I'm interested in a job at NASA. What should I do?" And the guy wrote back right away and asked me to send a résumé. I immediately got multiple additional postdoc offers, though I was not interested in doing a second postdoc. But then, a month or so later, I got an email saying, "There are these series of jobs coming up in Ames Research Center," which was the center for astrobiology. So I applied for the job and was hired. Then, within a month of getting there, I was walking around after Barry Blumberg, helping him start the Astrobiology Institute.

JJ: When was this?

CC: 1999.

JJ: Oh, so just a couple of years after the Allan Hills meteorite [ALH84001] discovery?

CC: Yes. I'm actually currently collaborating with Andrew Steele, who was on the original papers. I'm a visiting scientist at the Carnegie Institute right now, and my current research is working on statistical frameworks to be able to assess the confidence in a claim of life detection. Steelie was on the original papers, and then he went and demonstrated that pretty much everything they had identified as a "sign of Mars life" is actually explainable as either contamination from Earth or nonbiological processes on Earth.

JJ: I'm so fascinated by that case. So many people think that we've already discovered life on Mars because they heard Clinton's speech about it. But they don't know about the controversy that followed.

CC: How knowledge is made is a really interesting process—the facts are only the beginning, and then there's community acceptance. The facts are almost irrelevant in some cases, well, particularly these days, to community acceptance of things. You have the facts, and then, once the facts are established, *then* there's community acceptance. And community acceptance can go up and then go down again on the same facts. Oftentimes, "We have concluded not," particularly for Viking, is actually, "We don't know."

And that's where my statistical work is very interesting because, on Viking, they had two experiments. They had the life-detection package that was looking for metabolism, and they had the mass spectrometer that was looking for chemistry—dead bodies. Gil Levin, the PI [principal investigator] of the life-detection package, still to this day is publishing papers claiming that his instrument found evidence of metabolism on Mars. But the mass spectrometer found evidence only of simple organic compounds, chlorinated methane and very, very, simple organic compounds. The Jet Propulsion Lab built the mass spec, and cleaned it extremely carefully to get rid of all the Earth contamination, like cleaning compounds. But, when they found chlorinated methane on Mars, they said, "Oh, that has to be cleaning fluid."

If you remember, the Phoenix mission also found evidence of perchlorates and simple chlorinated compounds on Mars. The SAM instrument on the Mars Science Laboratory (MSL) mission, despite looking through all the Earth contamination on the Curiosity rover, has found evidence of simple chlorinated compounds that MSL have demonstrated are a result of reaction of organic carbon in the Mars soil with perchlorate in the Mars soil.

On Viking, we have a metabolism instrument that claims to have found life, *still* claims to have found life, and a mass spec., the results of which were misinterpreted. The interpretation of those results was not correct—this has been demonstrated, because people on the SAM instrument went back to look at the Viking traces, and they found evidence of the same compounds that SAM identified as from Mars in the traces from Viking.

JJ: Oh, wow.

CC: So, on the result of one positive experiment and one negative experiment, what should be the interpretation of those data? If you've got one positive and one negative result, what does that tell you?

JJ: It tells you that we don't know.

CC: Bingo. But the conclusion from Viking was that *there's no life on Mars*. And, for that reason, there was not Mars exploration by either the US or the Soviet Union for the subsequent twenty years. And yet, the metabolism experiment measured something accurately. And the mass spec. interpretation was wrong.

JJ: That tells you so much about—

CC: How science works.

JJ: And also public understanding of how science works, right? Anything other than a resounding "yes" has got to be "no"—there's no in between.

CC: Right. But I would say, even within the scientific community, there's a major—and this is getting back to my other experiences at NASA—a major problem with how people who want to explore Mars report results. And, of course, then public understanding is related to how the things get reported. So, the public won't have an accurate understanding of things they don't get told accurately. But the issue with planetary protection goes back even prior to Viking. People building spacecraft don't want there to be life on other planets, because if there is life on other planets, then they won't be able to drive their dune buggies all over Mars. It goes back beyond that to the Ranger program that first explored the moon. There was a huge argument between NASA Headquarters and Jet Propulsion Lab as to whether they should, for planetary protection purposes, have to bake their spacecraft.

Joshua Lederberg is one of the people who's weighing in on this at this time, and Carl Sagan was involved. In the late 1950s, early 1960s, this was when planetary protection was being developed, in the WESTEX report and internationally. Spacecraft engineers just didn't want to have to bake their spacecraft. They didn't want to have to clean spacecraft because they had ignored concerns about biological contamination in designing the hardware, and they still ignore them. To this day, there are no requirements in the spacecraft design constraints that Mars instrument designers get to have their hardware be resistant to sterilizing heat temperatures.

If spacecraft are going to go land on Mercury or Venus, they have to tolerate heat because that's an environmental operating condition. But there are no requirements that the instruments being designed for Mars missions ever have to consider the possibility of being heat-sterilized. So this has created the situation of adding planetary protection after the fact, and this causes the engineers to be resistant to implementation, because they never included it in their design constraints. So, of course it's going to be a problem. That's a cultural problem, not a scientific or engineering problem.

JJ: Sterilization seems directly tied to preserving the purity of scientific inquiry. You don't want to find the life you brought with you, right? You've been so clear about this over the years.

CC: Biology. It has to do with biology. If you don't care about biology, you just want to make this go away.

JJ: That's fascinating. I write about the impact that *The Andromeda Strain* had on the Apollo mission and the way that it provoked all of this public concern. But at the same time, so many scientists themselves were weighing in on it and writing letters to NASA saying, "This is ridiculous." They just thought all the precautions were absolutely unnecessary. And Lederberg says, "Well, it's unlikely that there's anything of concern on the moon, but it's still good practice for when we're going to go elsewhere, like Mars."

CC: Yes. And actually, the policy and review systems that were set in place for the Apollo program are still what should be done on human exploration. But it's important to look at where the people getting involved came from. Look at their scientific background. Look at their funding sources. It's very important to understand what the agendas are. It's kind of like tobacco and sugar. They're trying to shift public discourse by what they do in the public sphere. But it's not from an unbiased starting point. So it's very important to understand those kinds of questions. What is the agenda? What are these people being funded for? What are they interested in doing? Why are they saying these things? How does it help them to say this? So it's sort of the "follow the money" question, but it's not only money. It can be other things too.

JJ: What do you think it would do for the life sciences if we were to find strong evidence of life off planet?

CC: It's an interesting thing. But it isn't paradigm-breaking in any way. It's just, "Okay. That gives us an *n* of 2 for understanding how life started." I mean, this whole idea of panspermia—that's one of the interesting questions of astrobiology: Is life on Mars related to life on Earth? And this is why planetary protection also is so important. I have had senior managers tell me, "Well, we don't have to worry about putting Earth life on Mars because we know how to deal with Earth life." How many organisms *not* from Earth have killed humans, in all of human history? And of all the life on Earth? So, if we have contamination from Earth and we want to see life on Mars and we don't even know if the life on Mars is related to us or not, these are important reasons for why contamination control is important. But the people in charge of doing this have already made up their minds that it's not important.

JJ: That was definitely one of Lederberg's points, too. It seems that he was really driven by the prospect that space exploration and the discovery of life—that it would make biology a more universal science, that biologists were limited if they only have an *n* of 1, only life on Earth.

CC: I would say that is actually an aspect of the scientific culture of Lederberg's time, the absence of a "theory of biology"—and this has changed. In particular, in biology, it has changed because of the genome projects and, in the minds of the people doing this work, our ability to say *everything*. They think we can now catalogue everything that makes up an organism. That's not true, of course, because they've overlooked the physical structures. The American Society for Cell Biology used to sell T-shirts with the motto *Omnis forma ex DNA*, "Everything is made from DNA." That's blatantly false. But molecular biology has so overtaken modern biology that there's this idea that having the cookbook—and, as we know from RNA editing, it's even an inaccurate cookbook! Having the cookbook lets you describe the banquet of all of life on Earth, which is mostly made not of DNA but cellulose and bones and teeth and all of the organic and inorganic compounds in an organism. The DNA may tell you how to make the proteins that will then build a cell, but it doesn't in any way tell you how to put it all together correctly. A cookbook doesn't tell you how to make dishes and silverware or set a banquet table.

My background is in microscopy. So, I look at cells, the structures, how the actin filaments put themselves together in order to create a muscle. That information is nowhere in the DNA. The structural information has been passed down, really descent from generations, like Darwin was saying, in how the proteins connect each other. If you have this whole bunch of sequences of DNA and you put them in a tube with ribosomes, you will never get an organism. Because the information on how those things get connected is embedded in the membrane structure and the positions of the proteins in the membrane and the positions of all the other compounds in a cell.

So, the structural information is a parallel transmission of information, which is nowhere characterized in the DNA. And most of modern biology, because it's based on molecular biology, has forgotten this. It's the same problem I mentioned before: "Neurospora crassa: Is that a plant or an animal?" Molecular biology is not all of biology. But it has so pervaded modern biology that it has satisfied Lederberg's concern about the absence of a theoretical framework.

JJ: Right. That makes so much sense. I hadn't considered DNA as a potential answer to Lederberg's question.

CC: Yes. So, in fact, a lot of the ways that people want to look for life on Mars is to look for DNA. And how are you going to find it if the organism doesn't have DNA? But the life-detection instrument from MIT was a DNA-based instrument. A lot of these people who are proposing to look for life, they are looking for Earth life. And it's a great set of instruments for planetary protection, but I guarantee you'll find what you're looking for.

JJ: That seems to be the key thing, and it seems related to the definitions people use, too.

CC: Yup, it's really easy to look for your keys under the lamppost. That doesn't mean that's where they are.

JJ: I love that. So, let's shift gears a bit from science to ethics. What kinds of ethical issues emerge in space exploration? What should be people be paying attention to?

CC: I am a biologist, so I'm interested in understanding what is there—whether there's indigenous life on Mars that is related to us or not related to us and then how that life has altered the environment. Because, of course, the entire environment on Earth, the planet, down to the core probably, has been altered by biology. Oxygenic photosynthesis has embedded far more oxygen in Earth rocks than normal, even down in the mantle of the Earth. If you took a meteorite from Earth, a nonbiological entity, you would be able to tell that the Earth had something weird going on. Because there's too much oxygen. There has been too much input solar energy transformed into chemical energy by the release of oxygen during oxygenic photosynthesis. Mars is not like that. This is one of the things that Andrew Steele is very effectively demonstrating.

Mars is a much more reduced planet. There has not been a transference of solar energy to chemical energy by doing photosynthesis on Mars. So that makes Mars interesting from a geological standpoint, as a planet that's rocky and somewhat similar to Earth but in the absence of this enormous amount of additional chemical energy captured from insolation. You can tell that Mars is not a biological planet at a gross level, because the materials that we get in the meteorites from Mars are not oxidized the same way that material from Earth would be oxidized. From a biology standpoint, it's interesting to understand what the environments are like with and without life. And, in order to do that, you have to explore carefully.

My ethics are embedded in my interest in knowledge and accurate measurement of environments. Also, I want to avoid the Heisenberg problem: I don't want to alter something when I'm observing it. If you don't care about biology and don't acknowledge the extent to which organisms can alter their environment, which most geologists do not, then this becomes an absence of a priority, because you think it won't happen. You just assume it won't happen. It's the same spacecraft engineering problem. People assume this won't be a problem. They in fact go out of their way to explain and justify why it won't be a problem, in contradiction to fact. "Oh, Earth organisms. We know how to deal with Earth organisms," says somebody whose colleague had their mother die of an infection in the hospital. It's a blind spot.

They're just not thinking about what they're saying, but they use that blind justification as a reason why they don't have to do stuff they don't want to do. And it fundamentally is an attitude problem. They keep claiming that it's a cost issue or a technology issue. But, in fact, as I said, they didn't include those design constraints into their requirements for instrument design. I have heard a senior manager in the Mars program say that he will invest $50 million in figuring out how to do planetary protection right, and then a new manager comes in and says, "No, I don't want to do this." So, it's fundamentally an attitude problem. If you believe it's not a problem, you will decide and come up with reasons why you don't have to think about it. And you will engineer the culture such that the culture opposes this.

SpaceX never got punished by the Federal Aviation Administration [FAA] for the fact that their launch certification application for the Tesla car was wrong. It was blatantly wrong, everybody knows, but there were no consequences. Tesla even admitted it: after other people started publishing that the trajectory was different than they'd reported, Tesla admitted that they'd "made a mistake" in their calculations. The "mistake" in their calculations had the result that the Tesla car reached exactly the orbit of Mars about two weeks away from when Mars would have been there.

JJ: Now that we're moving into the age of private space exploration—does the Outer Space Treaty still apply?

CC: The US is a signatory to the treaty. The treaty says that the treaty constraints apply both to government and nongovernmental entities of treaty signatories. The Committee on Space Research advises the UN and provides guidelines that have been recognized by UN COPUOS [Committee on the Peaceful Use of Outer Space] as being consistent with compliance with Article IX of the Outer Space Treaty. As PPO, I was trying to get the FAA and the State Department and the rest of the federal government all to work together and create a government-wide Office of Planetary Protection that could be the necessary oversight and resource for commercial space. Look at the Trump administration; look at commercial spaceflight. You have congresspeople saying that we were going to put humans on Mars within the next five years. How are you going to comply with the Outer Space Treaty? Article IX, specifically? Again, it's the cultural attitude of, "We don't believe this is a problem. Therefore, we're going to ignore it." If you want to understand the problem, do you know the book *1493*, by Charles Mann?

JJ: Yeah.

CC: The Columbian Expansion. They want to continue the same level of ignorance that was displayed during the Columbian Expansion. They don't care. They believe it's not a problem. They believe it hasn't been a problem on Earth, smallpox to the contrary, malaria to the contrary, $5 billion a year to the economies of Central America to the contrary. They just have decided it's not a problem, and therefore, they're going to ignore it. In space exploration, the people in positions of power are mostly engineers and not scientists. It's a cultural problem. They don't want to deal with it. And so they're going to ignore it. And then they're going to contaminate. And you may even have people like Elon Musk, who's going to contaminate deliberately and then claim, "Oh, we've already thrown one match into the forest. That means we can throw as many matches as we want to."

JJ: Yeah. Well, especially when you have for-profit enterprise, and it's so expensive to sterilize.

CC: It's not actually that expensive. Unfortunately, it's the classic set of attitudes that lead institutions to only satisfy the concerns of the small group of people in power. And all the concerns of everybody else, including the rest of society, get ignored.

JJ: Carl Sagan famously had such a romantic, idealistic way of thinking about the space program. In my chapter on Mars, I write about his line from *Cosmos* that we should leave Mars alone if we find life there. I'm sure you know this line well. What do you think about that idea?

CC: From a biological standpoint, I would want to leave it alone and observe it in a way that didn't harm it, avoid the Heisenberg problem. The foundation of biology is natural history. You observe things, but you don't disturb them. That's how you understand what they are, in themselves. Of course, the Columbian Expansion is the prime example of how limitations and careful investigation go out the window when people want to make money off something. My ancestors got to what is now the US in 1610. So, yeah, my ancestors were part of that problem. Within human culture, it's the dichotomy between wanting to understand things from an intrinsic "this is valuable information" standpoint and wanting to profit from them.

JJ: A classic ethical tension.

CC: Right. It comes down to a problem of, "Do I have enough resources already that I can afford not to take advantage of anything I encounter?" And it's a problem that science regularly deals with. Archaeology is one of the fields that has grappled with this pretty well, because there are only so many archaeological sites. And there's a choice to leave some of them to be investigated later, when you have better technology, that took a while for archaeologists to figure out. But they did eventually get there.

There are now sites, even parts of Pompeii, I think, protected from investigation so that future generations have the chance to go and use future—better—tools and learn more. And it's something that archaeology as a community had to come up with. Space exploration is not there. But planetary protection started that way. Planetary protection is probably the first time in human history that people consciously decided, "We will avoid making mistakes before we have the opportunity to make them." Planetary protection was discussed in 1957 at some of the very early International Astronautical Federation discussions—even before the launch of Sputnik. And that held only until humans could make money off it. As soon as humans can make money off something, then all of those pure and admirable ethical, high-moral-standard issues disappear, because somebody's going to make money. It's particularly problematic when you have a country like the US where the companies have so much political power.

There's a short story by Isaac Asimov about political revolution. The title is "In a Good Cause," and the line goes on, "there are no failures; there are only delayed successes." I loved this story from the time I first read it as a kid. Well, the problem with planetary protection is that there are no successes: there are only delayed failures. And we are currently seeing how the failure will no longer be delayed. Without a much stronger set of priorities placed on avoiding contamination, it will happen. The attitude of the commercial space side is, "We're not even going to try, because of course it will eventually happen." Which always makes me wonder whether they brush their teeth every night.

Conclusion | *De Anima*

One of my favorite descriptions of life is found in Thomas Mann's *The Magic Mountain*. The novel follows the journey of the protagonist, Hans Castorp, through his seven-year stay at a tuberculosis sanitarium high in the Swiss Alps, and it features frequent meditations on life, sickness, death, and the mortal body that joins them. In one such passage, Mann describes Castorp burying himself in books about biology, anatomy, and physiology during "frosty nights" in which he bundled in wool and fur and "read with burning interest about life and its sacred, yet impure mystery."[1] In pages of abject description of the human body that oscillate between desire and repulsion, the living body is described as a process of exchange, "nourishment sucked in and excreted, an exhalation of carbon dioxide and other foul impurities of a mysterious origin and nature." "What was life, really?" Mann asks three times in as many pages, and he answers with a familiar dodge: "no one knew." It was not "matter," he writes; "it was not spirit. It was something in between the two, a phenomenon borne by matter, like the rainbow above a waterfall, like a flame."[2]

What I have hoped to show in this book is that whatever else it might be, really, this phenomenon we call life is a rhetorical one. We have been wandering around the neighborhood of *What is life?* for many pages, and here at the end, I find myself wanting to give you an answer that offers a way out of this aporia, an answer that, as it so happens, would also underscore the importance of my professional way of looking at the world. I find myself wanting to say that vital advocacy serves a constitutive role, that these rhetorical actions on life's behalf are, at long last, what bring *life itself* into being. I find myself on the verge of arguing for what we might call a rhetorical vitalism. But I pull back. A rhetorical perspective does lead us out of *What is life?* but not by answering it. Rather, it changes the terms of the question, encouraging us to examine what the invocation of life itself *does*: what it gathers together, what it divides, what it brings into being, and what it makes possible. To that end, I want to close with a final example of vital advocacy that inspires me, rooted in the breathing body that so captivated Hans Castorp in *The Magic Mountain*.

As I was bringing this book to a close, I attended a virtual talk by the political theorist Achille Mbembe at the end of the second spring of the COVID-19 pandemic. I could think of no one better to explain our grave global predicament, the sacrificing of "essential" workers, elders, and disabled people in the name of the economy, the unequal distribution of vaccines and death, so much death, striated across all-too-familiar racial, economic, and global lines. But while Mbembe had plenty to say about the necropolitical machinations of the pandemic, I found myself struck by how often the world's most eminent theorist of death returned to the subject of life.[3] Theories of biopolitics illustrate how life itself can be wielded to control human bodies and manage populations, and theories of necropolitics reveal how violence and death are often enacted on life's behalf.[4] But, as Michelle Murphy writes, it is also possible to imagine a politics of life beyond biopolitics, a politics of collective life that can serve as both a means of critique and a mode of resistance.[5] And *this* is what I was hearing in Mbembe's talk: life deployed not as a tool of control from the top down but as the stuff of solidarity from the bottom up, the grounds for a radical vital politics.[6] I use "vital politics" here in the sense that Nikolas Rose means it, as a "politics of life itself," but with a crucial difference. While for Rose vital politics concerns "our growing capacities to control, manage, engineer, reshape, and modulate the very vital capacities of human beings as living creatures," the vital politics we see in Mbembe invokes life itself for its potential to relate, repair, and rebuild a new world in common across species.[7]

This perspective can best be seen in "The Universal Right to Breathe," a brief essay that Mbembe published in the terrible spring of 2020. Though the pandemic had brought the world to a halt, it is but one "spectacular expression" of many simultaneous crises, Mbembe argues, each of which is rooted in the alienation of humanity from the living world. Right now we are at war, a war that modernity declared on life itself long ago, even if some of us haven't realized it yet, even while some of us serve as its foot soldiers and others its targets.[8] Caught in the fog of the present moment, we are on the verge of "something that still eludes our grasp," an inflection point marked by pervasive injustice and inequity, waning democracy and waxing nationalism, life constricted by borders, gates, and screens, all while the threat of ecological collapse looms: interlocking crises created and powered by a predatory, extractive global economic system that profits from—indeed, was built by, powered by—misery and death. The threat from these crises is "increasingly existential," Mbembe states bluntly.[9]

Whether it takes shape as the hypoxia of COVID-19, the choking of the planet in ever-increasing levels of carbon dioxide, or "everything that, in the long reign of capitalism, has constrained entire segments of the world population, entire races, to a difficult, panting breath and life of oppression," the war on life is a war on breath. To emerge on the other side of this war, it will therefore be necessary to reimagine what breathing is, what breathing does, and what breathing means; we must think of breathing beyond "its purely biological aspect and instead as that which we hold in common." Mbembe transforms breath from a biological function to a political necessity: "a right that belongs to the universal community of earthly inhabitants, human and other." The pandemic reveals the world at a point of juncture, a collapse in the fiction of unlimited human agency, a "planetary impasse" that offers a moment of pause, not for escape but for a recalibration of the relations among "human life and planetary life." The task before us thus "is no less than reconstructing a habitable earth to give all of us the breath of life. . . . Are we capable of rediscovering that each of us belongs to the same species, that we have an indivisible bond with all life? Perhaps that is the question—the very last—before we draw our last dying breath."[10]

As I argued in chapter 3, species thinking can be a problem. Invocations of the human species often proclaim unity in a way that erases difference, occluding the effects of injustice and eliding history and, by extension, responsibility. At its very worst, species thinking uses difference as a means of violent demarcation, unifying some under the name of human by casting others outside its protective borders. But here Mbembe uses the idea of species in a different way—an invocation of species *across* species. The "each of us" in the preceding passage is not a human us, or rather not *just* a human us, but a living us. Species is often defined by an individual's capacity to reproduce itself. But this kind of species is defined by a collective responsibility to ensure the reproduction of life on Earth. This is a utopian vision, to be sure, but Mbembe—one of our most profound critics of colonialism and anti-Black racism—can hardly be called naïve. The future community he imagines will be built under the long shadow of capitalism, and on the ruins of the colony, the plantation, and the slave ship. The vital politics he imagines looks forward, but always with one eye on the violence of the past and present.[11] The future, he states, "cannot come at the expense of some, always the same ones."[12] What Mbembe is calling for—a joining *across* difference, united by the common fact of *living together*—is, in other words, what I have described in this book as "bioplurality."

"Breath" is often used as a synonym for "life." In many languages, the words for "breath" and "life" are the same, even as their referends differ. In Latin, for example, *anima* means "life," "breath," and "air," as well as the ineffables that push us into the territory of vitalism: spirit, soul, force. As we have seen in this book, it is common to speak of life as a noun—*a* life, life itself. But one would never say "breath itself." Even to speak of *a* breath trips the tongue; it calls attention to itself; it stops time. *A* breath is a snapshot of a body in motion for which *before* and *after* are necessary context, except, of course, in two instances: the first breath and the last, which mark the boundaries between life, death, and the unknown. The moment of death is the moment when breath becomes air, as Paul Kalanithi so beautifully put it in his book of the same name. Maybe breath, like life, is thus better understood like a verb, as Dorion Sagan recommends. Breath does not exist outside the act of breathing, which is a biological function, to be sure, but as Mbembe reminds us, it is also a capacity that can be enabled or disabled across relations of power.[13]

All living things breathe, not necessarily with lungs. Insects take in oxygen through very small openings in their bodies called "spiracles." Plants inhale carbon dioxide, and they exhale oxygen in the act of photosynthesis. Even archaea and bacteria breathe. Some bacteria, like *Shewanella oneidensis*, found throughout the soil, even breathe in metals.[14] All living things breathe, not necessarily with air. "Each breath connects us to the rest of the biosphere," Sagan writes with Lynn Margulis, "which also 'breathes,' albeit at a slower pace."[15] Oxygen, nitrogen, and carbon dioxide fluctuate in the atmosphere by day, by season, by year, by epoch: respiration that pulses across the slowly spinning planet in rhythms known and unknown. "As long as we are alive," writes Sara DiCaglio, "we breathe together."[16]

In the first few pages of this book, I noted Robin Wall Kimmerer's point that "action on behalf of life transforms." What Mbembe suggests is that this action—what I have called in this book "vital advocacy"—has the potential to change no less than *everything*. Is that too bold a claim? Maybe. But I still believe it. If it is a time of big questions, as Mbembe argues, then it is also a time for big answers.[17] If what is required at this moment of planetary impasse is nothing less than rebuilding a habitable Earth—and that is what is required—then changing everything is precisely what needs to be done. Changing everything means rethinking the universal assumptions of the Western world and repairing the violent effects those assumptions have wrought over generations. It means rethinking land and water use, and it also means returning land to Indigenous people. It means rethinking our political and economic systems so that they benefit not just every human but also the nonhumans with whom we live in a relation of mutual responsibility, and the Earth that makes each of our lives possible. It means rethinking our unsustainable culture of production, consumption, and waste as well as the extractive systems of energy that enable and sustain it. It means rethinking the material, ethical, and political relations among human life and other forms of life (and not-life). For many, this involves reconsidering our understanding of "the" human, and reimagining humanity's place in the world, but also expanding our understanding of notions of life and life itself—not as something we have or even something we share but as something we do, and do together. Maybe this thing we call life exists outside rhetoric, and maybe it does not. *I am not going to answer this question*. But what I do know is that the movements for solidarity, justice, and repair necessary to ensure the breath of life for a we that is yet to be is one that will be made with rhetoric, even if not always with words.

Notes

Introduction

1. Haldane, *What Is Life?*, 58.

2. Bennett, *Vibrant Matter*, 62–81. For a history of the way that "life" came to be constituted in abortion discourse in the nineteenth century, see Stormer, *Articulating Life's Memory*.

3. Black Lives Matter (BLM) is probably the most visible political movement "on behalf of life" in recent years. However, BLM is not a movement on behalf of life itself or of human life as such but on behalf of *Black life*, and preserving that distinction is important. Because this book centers on life itself as it takes shape in advocacy across species, I am wary of using BLM or related political movements as a case study to avoid flattening their specific political purpose, mindful of Christina Sharpe's point that "the suffering of Black people cannot be analogized," and "'we' are not all claimed by life in the same way" (*In the Wake*, 63). For more on BLM, see K. Taylor, *From #BlackLives Matter to Black Liberation*. For a rhetorical perspective on how anti-Black violence has served as a form of civic engagement in the United States, see Ore, *Lynching*.

4. Foucault, *Order of Things*, 265, quoted in Doyle, *On Beyond Living*, 11. I am taking some liberties with this quote. Foucault is describing the process by which life emerged in the eighteenth century as a "transcendental" epistemological object (*Order of Things*, 244).

5. Doyle, *On Beyond Living*, 12. Here Doyle is expanding on Foucault, *Order of Things*, 278.

6. Doyle, *On Beyond Living*, 1. In a related point, Stephen Toulmin argues that there is "a genuine piety toward creatures of other kinds" produced by seeing them as ends rather than means, enabled by "cosmological sense of things" (*Return to Cosmology*, 272).

7. Kimmerer, *Braiding Sweetgrass*, 339.

8. The precise location of the boundary between the Earth and outer space is still a matter of controversy. See Goedhart, *Neverending Dispute*.

9. George, "Yuri Gagarin."

10. Gagarin, "Yuri Gagarin's First Speech."

11. White, *Overview Effect*.

12. Quoted in Planetary Collective, *Overview*.

13. For more on the rhetorical power of this sentiment, see Cosgrove, "Contested Global Visions"; Lazier, "Earthrise"; Welter, "From Disc to Sphere"; J. DiCaglio, *Scale Theory*; and Messeri, *Placing Outer Space*.

14. Boes, "Beyond Whole Earth," 157.

15. Ibid., 159–60.

16. Ferreira, "Seeing Earth from Space."

17. S. Chen, *Documentary on Shenzhou-9*, 288, quoted in Yaden et al., "Overview Effect," 3.

18. Howell, "James Irwin."

19. Schumacher Center for a New Economics, "Lindisfarne Tapes," accessed July 1, 2020, https://centerforneweconomics.org/envision/legacy/lindisfarne-tapes.

20. Quoted in Oliver, *To Touch the Face*, 115.

21. Quoted in French and Burgess, *In the Shadow*, 363.

22. Russell Schweickart, speech to the Lindisfarne Association, 1974, http://archive.org/details/RussellSchweickartA2 (my transcription).

23. Brand, *Space Colonies*, 110.

24. While the words sound similar, what I am calling "bioidentification" is very different from "biocertification," Ellen Samuels's term for the process of definitively identifying bodies according to race, gender, or ability. For more, see Samuels, *Fantasies of Identification*.

25. K. Burke, *Rhetoric*, 55.

26. See Sowards, "Identification Through Orangutans," for an example of how humans can be identified with nonhuman animals.

27. K. Burke, *Rhetoric*, 54; see also Mangold and Goehring, "Identification by Transitive Property."

28. K. Burke describes "a general *body of identifications* that owe their convincingness much more to trivial repetition and dull daily reinforcement than to exceptional rhetorical skill" (*Rhetoric*, 26; original emphasis). For a discussion of rhetorical accumulation, see Olson, *American Magnitude*, 27–68. I am drawing the language of "stickiness" here from Sara Ahmed, who uses the term throughout *Cultural Politics*.

29. K. Burke, *Rhetoric*, 31.

30. Kant, *Critique of Judgment*, 94.

31. Gunn and Beard, "On the Apocalyptic Sublime," 275.

32. J. DiCaglio, *Scale Theory*, 238.

33. E. Burke, *Philosophical Enquiry*.

34. Hitt, "Toward an Ecological Sublime," 609.

35. K. Burke, *Rhetoric*, 31. In this book, I use "substance" in this looser, more abstract way found in Burke's notion of consubstantiality. For a sustained discussion of how the understanding of life in theories of vitalism shifted from substance to force and complexity, see Hawk, *Counter-History*, 121–65.

36. K. Burke, *Rhetoric*, 20–21. Emphasis in original unless otherwise noted.

37. K. Burke, *Grammar*, 23.

38. Bryant, *Democracy of Objects*, sec. 2.3.

39. K. Burke, *Grammar*, 56–57.

40. Ibid.

41. Durham, "Burke's Concept of Substance," 358.

42. Doyle, *On Beyond Living*, 1.

43. Mitchell, *Experimental Life*, 1.

44. Lyne, "Bio-rhetorics"; Keränen, "Addressing the Epidemic of Epidemics."

45. Notable exceptions are Doyle, *On Beyond Living*; Doyle, *Wetwares*; Hawk, *Counter-History*; Rowland, *Zoetropes*; Stormer, *Articulating Life's Memory*.

46. Here I am thinking specifically of Rowland's concept of "zoerhetoric," which describes not a general rhetoric of life itself but a specific manifestation in "public discourses or practices that transvalue the status of a group of existents" across the lines of race, gender, species, or ability (*Zoetropes*, 16).

47. Hawk, *Counter-History*, 4.

48. Wilson, *Biophilia*, 1–2. Wilson, ever the evolutionary biologist, speculates that biophilia is an innate trait of organisms; however, there is no empirical evidence for this idea. See Simaika and Samways, "Biophilia."

49. Wilson, *Biophilia*, 138.

50. Kimmerer, *Braiding Sweetgrass*, 49.

51. Ibid., 58.

52. Mignolo, *Darker Side*, 119.

53. In this book, I use the term "Indigenous" as a political analytic, following Maile Arvin (Kanaka Maoli). As a category of identity, it can risk flattening the specifics of particular Indigenous peoples, even as it enables political solidarity among them. But as a political analytic, "Indigenous" is critically useful a term "in articulation with raciality and coloniality" that "refers to the historical and contemporary effects of colonial and anticolonial desires related to a certain land or territory and the various displacements of that place's original or longtime inhabitants" (Arvin, "Analytics of Indigeneity," 121). See also Na'puti, "Speaking of Indigeneity," 496.

54. DiCaglio, Barlow, and Johnson, "Rhetorical Recommendations," 446.

55. Mitchell, *Experimental Life*, 2.

56. S. Murray, "Aporia," 12–13.

57. Derrida, *Aporias*, 3, cited in S. Murray, "Aporia," 8.

A Conversation with Dorion Sagan

1. Margulis and Sagan, *What Is Life?*, 14.

2. See, for example, Vernadsky, *Biosphere*.

3. Jeffers, "Treasure," 100.

4. Margulis and Sagan, *What Is Life?*, 20.

Chapter 1

1. Dalrymple, "Pipeline Route Plan."

2. The Oceti Sakowin (the people of the Seven Council Fires) are a confederacy of the Dakota, Lakota, and Nakota peoples also known as the Great Sioux Nation. For an account of the origin and development of the Oceti Sakowin and their systems of governance, the complex history of the term "Sioux," and their relationship to Standing Rock, see Howe and Young, "Mnisose."

3. Phyllis Young, "Standing Rock Speech to DAPL Company," Earthjustice YouTube video, 8:00, December 13, 2016, https://www.youtube.com/watch?v=_wlRdkP3Q7o.

4. Estes and Dhillon, prologue to *Standing with Standing Rock*, ix.

5. Castro-Gómez, *La hybris del punto cero*, quoted in Mignolo, *Darker Side*, xxxiii.

6. Mignolo, *Darker Side*, 80.

7. Wanzer, "Delinking Rhetoric," 650–51; see also Shome, "Postcolonial Interventions," 49.

8. Wanzer, "Delinking Rhetoric," 653.

9. Na'puti, "Speaking of Indigeneity," 496.

10. Biesecker, *Addressing Postmodernity*, 32–33.

11. Halévy, *The Growth of Philosophical Radicalism*, cited in Lukes, *Individualism*, 17.

12. Fanon, *Wretched of the Earth*, 11.

13. Foss and Griffin, "Feminist Perspective," 345.

14. Garland-Thomson, "Misfits," 593.

15. Burkhart, *Indigenizing Philosophy*, xix. While I find Burkhart's notion of ill fit a productive one, there is a broader conversation about the idea that Indigenous and Western knowledges are *incommensurable* that I haven't the space to address here. For more on this point, see Watson and Huntington, "They're Here."

16. Quoted in Sackey et al., "Perspectives," 395.

17. Olson, *American Magnitude*, 188.

18. Davis, *Inessential Solidarity*, 19.
19. K. Burke, *Rhetoric*, 20–21.
20. Ibid., 22.
21. Davis, "Burke and Freud," 128.
22. Foss and Griffin, "Feminist Perspective," 338–39 (my emphasis).
23. Querejazu, "Encountering the Pluriverse."
24. Foss and Griffin, "Feminist Perspective," 343.
25. K. Burke, *Rhetoric*, 115, 164.
26. K. Burke, *Grammar*, 670.
27. K. Burke, "Methodological Repression," 404, quoted in Davis, *Inessential Solidarity*, 23.
28. Biesecker, *Addressing Postmodernity*, 46.
29. Davis, *Inessential Solidarity*, 22. For more on the way that the living body structures Burke's thought, see Hawhee, *Moving Bodies*, and especially pages 30–54, where she describes his engagement with the vital mysticism / mystic vitalism of William James.
30. Gilbert, Sapp, and Tauber, "Symbiotic View," 326.
31. Querejazu, "Encountering the Pluriverse."
32. C. Taylor, *Sources of the Self*, cited in Gilbert, Sapp, and Tauber, "Symbiotic View," 326.
33. I am inspired here by the work of Lynn Nyhart and Scott Lidgard on the history of the biological individual, specifically their article "Individuals at the Center of Biology." My deep thanks to Lynn for our conversations on the topic of the biological individual, which inspired some of my thinking in this chapter.
34. Lidgard and Nyhart, introduction to *Biological Individuality*, 4.
35. For a further elaboration of this point, see J. DiCaglio, *Scale Theory*, 147–82.
36. On this point, see Peters, "My Body, My Cells."
37. J. DiCaglio, *Scale Theory*, 122–27.
38. Gilbert, Sapp, and Tauber, "Symbiotic View," 336, 326. For the sake of clarity, I have reversed the chronological order in which these sentences appear in the original.
39. Bedau and Cleland, introduction to *Nature of Life*, xxi–xxii.
40. Machery, "Why I Stopped Worrying."
41. For more on these approaches, see Mariscal and Doolittle, "Life and Life Only," which outlines several definitions and taxonomies of life in detail.
42. Jonas, *Phenomenon of Life*, 79.
43. Quoted in Canguilhem, *Knowledge of Life*, 69.
44. Ibid., 70.
45. Walsh, *Organisms*, 12; Walsh, "Objectcy and Agency," 167.
46. In "Vitalism," Benton argues that it is difficult to speak of vitalism at any time as a unified system and proposes instead categories of vitalism, such as "realist vitalism," "nomological vitalism," and the like. For a rhetorical history of vitalism that parses out some of these distinctions, see Hawk, *Counter-History*, esp. 121–65. Even further, as Osborne observes, within contemporary science, the line between these two perspectives has almost disappeared, with the advent of nano-engineering, synthetic life, and gene editing, "as if mechanisms had become vital and life mechanistic" ("Vitalism as Pathos," 191). For more on vitalism from a rhetorical perspective, see Doyle's *On Beyond Living* and especially *Wetwares*, which complicate and extend this story into the "postvital" present and future.
47. Hawhee points out that Aristotle's living soul is further split between the functions of judgment, a combination of intellect and sensation, and the capacity for exciting movement in space. The former he names as the "first characteristic of an animal" (*Rhetoric in Tooth*, 21).

48. Aristotle, *De Anima*, 24, 413a23–a26. Intriguingly, Aristotle seems to abandon the effort to define life itself, "an account which is not distinctive to anything which exists, which does not correspond to any proper and indivisible species, all the while neglecting what is distinctive. Consequently, one must ask individually what the soul of each is, for example, what the soul of a plant is, and what the soul of a man or a beast is" (28, 414b).

49. Nealon, *Plant Theory*, 39; see 29–48 for a fascinating discussion of the role of plant life in Aristotle and Heidegger.

50. M. Chen, *Animacies*, 4–5.

51. TallBear, "Beyond the Life/Not-Life Binary," 191. TallBear is extending Deloria's term from "American Indian Metaphysics."

52. Quoted in TallBear, "Beyond the Life/Not-Life Binary," 191.

53. Ibid., 190.

54. TallBear, "Indigenous Reflection," 233.

55. Watts writes, "In an epistemological-ontological frame, Indigenous cosmologies would be examples of a symbolic interconnectedness—an abstraction of a moral code. It would be a way in which to view the world—the basis for an epistemological stance" ("Indigenous Place-Thought," 21).

56. TallBear, "Beyond the Life/Not-Life Binary," 198.

57. Burkhart, *Indigenizing Philosophy*, 194.

58. Foss and Griffin, "Feminist Perspective," 344.

59. Burkhart's playful textual alter ego, the Lakota spider-trickster Iktomi, suggests that "the Western thinker" who believes that stones have no language has obviously never spoken with "Inyan, the grandfather stone," who "would likely have his feelings hurt" to hear that he did not have interests or that he could not be harmed, "and then things could get real messy in the next sweat lodge. . . . If Inyan did not feel pleasure or pain then why did he choose to cut himself open to release his blue blood, the water that became the lifeblood of the earth and was responsible for the generation of human beings in the first place?" (*Indigenizing Philosophy*, 172).

60. Grant, "Writing 'Wakan,'" 71.

61. Charland, "Constitutive Rhetoric," 139.

62. Olson, *Constitutive Visions*, 28.

63. Hodges, "Yes, We Can."

64. Metildi, "Water Is Life."

65. Here I'm slightly rephrasing the well-known description of terministic screens found in K. Burke, *Language as Symbolic Action*, 45.

66. Burkhart, *Indigenizing Philosophy*, 71–72.

67. Estes and Dhillon, introduction to *Standing with Standing Rock*, 2–3.

68. Burkhart, "Be as Strong," 28.

69. Vitalist International, Twitter post, July 1, 2020, 3:52 p.m., https://twitter.com/VitalistInt/status/1278431283547459589.

70. Vitalist International, "Life Finds a Way." The manifesto seems to be authored by the "Atlanta Faction" of the Vitalist International, but since it speaks of the purpose of VI in general terms, I have chosen to analyze it that way.

71. Ibid.

72. For more on this term, see Mack and Na'puti, "'Our Bodies.'"

73. Vitalist International, "Life Finds a Way."

74. Estes and Dhillon, introduction to *Standing with Standing Rock*, 3; see also American Horse, "We Are Protectors."

75. Montana Journalism, "Dallas Goldtooth, Protector vs. Protestor," Vimeo video, 1:07, accessed May 20, 2020, https://vimeo.com/194079656.

A Conversation with Kyle Whyte

1. See Whyte, "Against Crisis Epistemology," 7.

Chapter 2

1. Keller, "Deep Ecology," 210.
2. DiCaglio, Barlow, and Johnson, "Rhetorical Recommendations," 446.
3. Foreman, *Confessions*, 175.
4. Oelschlaeger, *Idea of Wilderness*, 304.
5. DiCaglio, Barlow, and Johnson, "Rhetorical Recommendations," 441.
6. Bookchin, "Crisis."
7. Næss, "'Man Apart' and Deep Ecology," 185.
8. Rowland, *Zoetropes*, 2–3.
9. Krabbe, "Arne Næss," 529.
10. In Scandinavia, *friluftsliv* has merged with Næss's thinking to such an extent that it has been described as "a way of living Arne Næss's Deep Ecology" (Gelter, "Friluftsliv," 79).
11. Henderson and Vikander, *Nature First*, 7; Gelter, "Friluftsliv," 79.
12. Næss, *Ecology of Wisdom*, 51.
13. Ibid., 45–64. This was in keeping with an old Norwegian tradition. Norway's old farms were given names that they kept for centuries, and the people who lived and worked there often took those names as their own. My maternal grandmother's surname, Tasa, for example, was the name of a farm in southern Norway, where generations of my ancestors lived and worked but which they did not own. They shared this name with other families who lived and worked there, even though they were not genetically related.
14. Murphy, *Economization of Life*, 138.
15. Næss would go on to speak favorably of *The Limits to Growth*, describing its beneficial impact on environmental policies and urging people in the deep ecology movement to engage with people and literature outside their communities. "Different methods of communication and different types of rhetoric will reach different kinds of people," Næss explained, "and this is essential if the movement is to be anything but a small partisan faction" (*Ecology, Community, and Lifestyle*, 153).
16. Anker, "Deep Ecology in Bucharest," 56–58.
17. Næss, "The Shallow and the Deep," 99.
18. Næss, *Ecology, Community, and Lifestyle*, 36.
19. Næss's original seven principles were (1) rejection of the man-in-environment image in favor of the relational, total field image, (2) biospherical egalitarianism, (3) principles of diversity and of symbiosis, (4) anticlass posture, (5) fight against pollution and resource depletion, (6) complexity, not complication, (7) local autonomy and decentralization ("The Shallow and the Deep," 95–98).
20. Sessions and Næss offered the principles to the readers of *Earth First!* with an invitation for rewording, additions, or deletions, but no changes seem to have been made.
21. Sessions and Næss, "Basic Principles of Deep Ecology," 19. All subsequent quotes from the principles come from this source.

22. Næss, "The Shallow and the Deep," 96.

23. Ibid., 97.

24. Anker, "Deep Ecology in Bucharest," 58.

25. Schweitzer, *Reverence for Life*; Goodpaster, "On Being Morally Considerable," 310; P. Taylor, *Respect for Nature*, 60–70; Callicott, "On the Intrinsic Value."

26. Diehm, "Identification with Nature," 2.

27. This idea is strikingly similar to ideas found in Indigenous ontologies, as Stan Wilson (Opaskwayak Cree) argues ("Self-as-Relationship," 91, cited in Clary-Lemon, *Planting the Anthropocene*, 64). According to Burkhardt's trickster alter ego Iktomi, "Arne Næss appropriated the Sherpa culture of reverence for the Himalayas. Iktomi thinks Næss 'transparently sensed' the reverence the Sherpas had for those mountains and extended through delocality this reverence to cover natural objects in general" (*Indigenizing Philosophy*, 171).

28. Næss, "Self-Realization," 227.

29. Fox, *Toward a Transpersonal Ecology*, 231 (my emphasis).

30. Ibid., xi.

31. Ibid., 249–50.

32. Ibid., 258.

33. Ibid., 253.

34. Diehm, "Identification with Nature," 3.

35. Leopold, *Sand County Almanac*, 192.

36. Diehm, "Identification with Nature," 3.

37. Ibid., 12–14 (my emphasis).

38. Reed, "Man Apart," 56.

39. Ibid., 63.

40. Plumwood, "Deep Ecology," 64, 65.

41. For more on the violence of residential schools from a rhetorical perspective, see Salas, "Decolonizing Exigency."

42. Plumwood, "Deep Ecology," 67.

43. Chávez, *Queer Migration Politics*, 125.

44. Næss, "Self-Realization," 227.

45. Diehm, "Identification with Nature," 12.

46. Næss, *Life's Philosophy*, 114.

47. Davis, *Inessential Solidarity*, 26.

48. Haraway, *Staying with the Trouble*, 103.

49. For an overview of ethical extension, see Nash, *Rights of Nature*, 3–12.

50. Rowland, *Zoetropes*. For more on the rhetorical connection between the moral standing of animals and people with disabilities, see, for example, Lewiecki-Wilson, "Ableist Rhetorics"; and J. Johnson, "Disability, Animals."

51. Diehm, "Identification with Nature," 15.

52. As Stacy Sowards points out, some other-than-human entities are easier for audiences to identify with than others ("Identification Through Orangutans," 58). There might be something else at work in this example with regard to the way that rhetorical magnitude is deployed—to *begin* with an insect (rather than, say, an orangutan), identification is easily extended to larger entities that typically tend to be granted more moral significance or a higher place on what Rowland calls the "zoerhetorical hierarchy" (*Zoetropes*, 14).

53. Næss, "Self-Realization," 227.

54. Nash, *Rights of Nature*, 4.

55. I am thinking here of Sara Ahmed's point regarding the intertwining of love and hate in white identity politics in *Cultural Politics*, 42–61.

56. Næss, *Ecology, Community, and Lifestyle*, 171–72.

57. This language is from Bergson, quoted in Bennett, *Vibrant Matter*, 2. Do stones, too, want to persist in their own being? Spinoza would say so, or at least falling stones do. Are they precarious? Not in Judith Butler's sense of the term, which is tied to injurability and killability. Do stones have faces? Not in the stretchiest application of Levinasian ethics, at least as he presented it. But, returning to the argument of chapter 1, do stones have "faces" in another framework? Absolutely. I think of the rock formation that US settlers call Mount Rushmore, for example, which the Lakota call Six Grandfathers or *Tȟuŋkášila Šákpe*. Tȟuŋkášila Šákpe has a face to the Lakota people but not to the settlers who *carved literal faces into it*. For more on the Inyan Oyate, see Burkhardt, "Be as Strong" and *Indigenizing Philosophy*.

58. Barnett, "Thinking Ecologically," 20.

59. See, for example, Davis, *Inessential Solidarity*, 144–66; Gehrke, "Ethical Importance"; Diehm, "Ethics and Natural History"; Derrida, *Animal*.

60. Butler, *Precarious Life*, 134 (my emphasis).

61. Butler, *Frames of War*, 7.

62. Ibid., 13–14.

63. Barnett, "Thinking Ecologically," 26.

64. Butler, *Frames of War*, 14.

65. Barnett, "Thinking Ecologically," 25–26.

66. Cox, "Die Is Cast," 230.

67. Barnett, "Thinking Ecologically," 37 (my emphasis).

68. Fox, *Toward a Transpersonal Ecology*, 257.

69. Ibid., 260.

70. The tree of life is a central, enduring metaphor in the biological sciences, a figure famously associated with Darwin's *Origin of Species*. For more on the history of the metaphor from a rhetorical perspective, see Gross, "Darwin's Diagram"; Miller and Hartzog, "Tree Thinking."

71. Fox, *Toward a Transpersonal Ecology*, 261.

72. Another way of thinking about how life works here is as a frame, which Barnett explains draws the rhetorical and ethical aspects of Butler's argument together. Frames, he writes, "perform the important work of delineating what does or does not come into view; they shape what can and cannot be sensed by audiences as reality. . . . Framing, then, is intimately related to response: since frames shape what will or can appear as reality, they also shape how audiences might respond. And in this sense, frames give rise to modes of responsibility" ("Thinking Ecologically," 35).

73. Fox, *Toward a Transpersonal Ecology*, 268.

74. Barnett, "Thinking Ecologically," 26.

75. Bookchin, "*Social Ecology Versus Deep Ecology*."

76. Reed, "Man Apart," 65. While Hardin enjoyed a long career as a respected ecologist, he also associated with a number of nativist, anti-immigration, and white nationalist groups and derided "multiethnic society" as a "disaster" and "insanity" (Southern Poverty Law Center, "Garrett Hardin," https://www.splcenter.org/fighting-hate/extremist-files/individual/garrett-hardin, accessed June 8, 2022). For a very different boat analogy, see Whyte, "Way Beyond the Lifeboat." In this essay, Whyte constructs a complex allegory of vessels that range from canoes (which represent Indigenous peoples) to aircraft carriers (which represent nation-states) to hovercraft (which represent corporations), each of which has a different relationship to the other vessels and the water that draws them together. The aircraft carriers and hovercraft, for

example, create turbulence in the water, yet the people in them, high above the water, see neither the water nor the trouble they are making for canoes. No matter how the aircraft carriers and hovercraft try to move closer to help the canoes, they destabilize the water even further. The problem, Whyte argues, is in the design. Whyte uses the allegory to argue that any attempt to address climate change must first address foundational issues like colonialism, capitalism, and industrialism: the "vanguard" work of Indigenous climate justice movements.

77. Bookchin and Foreman, *Defending the Earth*, 21.

78. Cagle, "Bees, Not Refugees." For more on Foreman's views on immigration, see his *Man Swarm*.

79. Abbey, "Immigration and Liberal Taboos," 43. Abbey calls this essay his "favorite" in the introduction to a collection of his work and describes its repeated rejection by mainstream newspapers and magazines as part of its "colorful history" (*One Life at a Time*, 2–3).

80. For more on conservation in Nazi Germany, see Uekotter, *The Green and the Brown*. For a wide-ranging critical history of the relationship between philosophical vitalism and racism, see Jones, *Racial Discourses*; for a narrower history of Ludwig Klages, "life philosophy," and its influence on Nazi ideology, see Lebovic, *Philosophy of Life and Death*.

81. Arktos, "Pentti Linkola: Radical Environmentalist and Deep Ecologist," https://arktos.com/people/pentti-linkola/.

82. See the reviews on Amazon at https://www.amazon.com/Can-Life-Prevail-Pentti-Linkola-ebook/product-reviews/B007USAVRM.

83. Linkola, *Can Life Prevail*, 51.

84. Ibid., 59.

85. Ibid., 50.

86. Ibid., 68.

87. See, for example, Marra and Santella, *Cat Wars*.

88. For a thorough history of this connection, see Coates, *American Perceptions*.

89. Subramaniam, "Aliens Have Landed," 29–30.

90. Linkola, *Can Life Prevail*, 72.

91. Owen, "Small Budapest Publishing House."

92. Linkola, *Can Life Prevail*, 74.

93. Ibid., 13.

94. Ibid., 51.

95. Ibid., 35.

96. Ibid., 74 (my emphasis).

97. Ibid., 75.

98. Linkola was more explicit in his other writing, arguing that the Second World War "freed the earth from the weight of tens of millions of over-nourished Europeans, 6 million of them by ideally painless means, without any damage to the environment" (quoted in Protopapadakis, "Environmental Ethics," 593). For another perspective on the way that life is subjected to economic calculus in contemporary discourse about population, see Murphy, *Economization of Life*.

99. As Susan Kollin writes, "The loss of nature experienced by Euro-Americans often becomes directed toward the racial Other, who in turn is made responsible for that loss, becoming a target of environmentalism's denigration and blame" (*Nature's State*, 140).

100. Næss, *Selected Works*, 10:95; for more on this point, see 93–101.

101. Barnett, "Thinking Ecologically," 26.

102. Sara DiCaglio argues that the COVID-19 pandemic reveals the threat/care relationship between self and other in very material ways ("Breathing in a Pandemic," 386).

Chapter 3

1. Barnosky et al., "Sixth Mass Extinction."
2. Ceballos et al., "Accelerated Species Losses."
3. De Vos et al., "Estimating"; Pimm et al., "Biodiversity of Species."
4. Ceballos, Ehrlich, and Dirzo, "Biological Annihilation."
5. Ibid., E6096.
6. Lind, "*Unbearable Loss*," 14.
7. Watts, "Stop Biodiversity Loss."
8. For more on the way that public attention is pulled toward some examples of extinction rather than others, see Barnett, "Naming."
9. Legagneux et al., "Our House Is Burning."
10. Rittel and Webber, "Dilemmas."
11. Ceballos, Ehrlich, and Dirzo, "Biological Annihilation," E6095.
12. For example, see Barkun, "Divided Apocalypse"; Buell, *From Apocalypse*; and Killingsworth and Palmer, "Millennial Ecology." Discourse on extinction is an exemplar of what Robert Cox has described as the "locus of the irreparable" ("Die Is Cast").
13. Clary-Lemon, *Planting the Anthropocene*, 5.
14. Lewis, "Toward a Self-Critical Environmentalism"; Meadows, "Chicken Little"; Feinberg and Willer, "Apocalypse Soon?"; Veldman, "Narrating the Environmental Apocalypse."
15. Chakrabarty, "Climate of History," 213.
16. Gilroy, *Postcolonial Melancholia*, 4.
17. Pravinchandra, "One Species," 27–28.
18. As of this writing, VHEMT's Facebook page has over twelve thousand followers.
19. This language was used in the collective letter of support for Extinction Rebellion, right before it launched on October 31, 2018. See Green et al., "Facts."
20. Nealon, *Plant Theory*, 120, 109.
21. Heise, *Imagining Extinction*, 86.
22. Arendt, *Human Condition*, 175.
23. As it relates to humans, the concept of species is deeply raced and gendered, a point especially evident in the foundational taxonomies of Carl Linnaeus, who separated *Homo sapiens* into four separate subspecies according to race (Schiebinger, *Nature's Body*, 52–55).
24. Pravinchandra, "One Species," 45.
25. In this chapter, I have focused on human plurality; however, one could also examine how plurality might be applied to nonhuman species. For a different meaning of plurality applied to the other-than-human world, see Zoe Todd's 2014 study of the "fish pluralities" of the Inuvialuit of Paulatuuq. "Pluralities" in this case refers to multiple ways of knowing and defining fish, which allows Todd to show how humans and animals might together be seen as active agents within the political and colonial processes of northern Canada. For an approach to transspecies and biotic/abiotic alliances from a rhetorical perspective, see Druschke and Rai, "Making Worlds."
26. Arendt, *Human Condition*, 7 (my emphasis).
27. Ibid., 8.
28. Benhabib, *Reluctant Modernism*, 109, expanding on Arendt, *Human Condition*, 175.
29. Arendt, *Human Condition*, 179.
30. Ibid., 52.
31. Vatter, "Natality and Biopolitics," 137.
32. Arendt, *Between Past and Future*, 156.
33. Arendt, *Human Condition*, 3.

34. Ibid. (my emphasis).

35. For example, see Adams, "At the Table," 11.

36. Arendt, *Human Condition*, 2.

37. This brings to mind Georges Canguilhem's point that life is *original*, the precondition of all thought (*Knowledge of Life*, xx).

38. See especially Benhabib, "Models of Public Space"; Bernasconi, "Double Face"; C. Johnson, "Reading Between the Lines"; Gines, "Hannah Arendt, Liberalism"; Gines, *Hannah Arendt*.

39. Wynter, "Unsettling." Wynter argues that the West created Man twice: the first, emerging from a theological understanding of the world, she calls "Man1," whose being was conditioned on the exclusion of the heathen; the second, emerging from a scientific understanding of the world, is "Man2," whose racialized others were those deemed biologically defective.

40. Katherine McKittrick argues that Wynter's "overall project can be identified as that of a *counterhumanism*—one now ecumenically 'made to the measure of the world'" (Wynter and McKittrick, "Unparalleled Catastrophe," 11).

41. Ibid., 16 (my emphasis). "If biocentrists are right," Wynter admits, "then everything I'm saying is wrong; but, if I am right, I cannot expect them to accept it easily" (17).

42. "About the Movement," accessed June 10, 2002, https://www.vhemt.org/aboutvhemt.htm.

43. *These Exit Times* (*TET*), no. 1 (1991): 1.

44. Ibid.

45. *TET*, no. 4 (1994): 8. The role that gender plays in VHEMT is complex. While VHEMT is populated by male and female Volunteers, women sometimes appear in VHEMT discourse as a stumbling block. For example, one Volunteer, "Jack O.," describes the women in his life as resolute procreators: his first wife "whined and cried" until he "gave in" and had children. His sons chose to follow in his footsteps ("their wives may have put the pressure on, but so far, the lads have held firm"), but the daughter he "couldn't influence" (*TET*, no. 2 [1992]: 11). A woman Volunteer also espouses this position, describing her own urge to reproduce as overpowering: even believing "in the complete extinction of our kind, still I felt an extraordinary, terrifying, ravenous, soul-shattering physical NEED to reproduce," a "lust" that "consumed" her for a number of years (10).

46. *TET*, no. 2 (1992): 1.

47. Ibid., 7.

48. Ibid., 5.

49. Ibid., 1.

50. Ibid., 5.

51. Ibid., 10. Because VHEMT is so conscious of the language it uses, one of the difficult things to make sense of is its humor. VHEMT's newsletter, website, and other materials feature cartoons, jokes, countless puns (such as Knight's pseudonym, a wordplay on "let's unite"), and a certain light-heartedness that leads one to think that Knight and his Volunteers can't *possibly* be serious. But humor does not negate VHEMT's message; rather, it appears to be employed to make the message more palatable, a rhetorical strategy that is "generally less threatening" than others and "can avoid negative feelings" (*TET*, no. 4 [1994]: 9).

52. *TET*, no. 1 (1991): 1.

53. *TET*, no. 2 (1992): 2.

54. *TET*, no. 4 (1994): 1.

55. Ibid., 9.

56. *TET*, no. 2 (1992): 18.

57. *TET*, no. 4 (1994): 3 (emphasis in original).

58. Ibid., 11.

59. *TET*, no. 2 (1992): 4 (my emphasis).

60. Ibid., 9 (my emphasis).

61. For example, see Bonilla-Silva, *Racism Without Racists*. In the same issue on *TET*, Knight addresses immigration restriction as a form of population control. Critiquing national borders as arbitrarily drawn and enforced at "gunpoint," Knight suggests that we might draw borders around wilderness preserves instead, restricting what he calls human "in-migration" into wild spaces (*TET*, no. 2 [1992]: 2). Once again, VHEMT appears to be presenting a more progressive view here, advocating for wild spaces instead of against migrants. However, just a few pages later, the reader is treated to a source guide to other "population-oriented organizations" (8), presumably to guide Volunteers to people who also care about life on the planet. Just a little digging, however, reveals that many of these organizations that advocate population control also advocate for immigration restriction and nationalism, such as "Population/Environment Balance," an anti-immigration group that argues, in an echo of chapter 2, that the US has reached its "carrying capacity." "All About Balance," accessed June 8, 2002, https://www.balance.org/about.html.

62. *TET*, no. 1 (1991): 7.

63. Like VHEMT, the GLF describes itself as a "concept" rather than an organization. Church of Euthanasia, "Gaia Liberation Front" (1994), accessed January 17, 2021, https://www.churchofeuthanasia.org/resources/glf/glfsop.html.

64. *TET*, no. 2 (1992): 2.

65. Haraway, *When Species Meet*, 11.

66. Rivers, "Deep Ambivalences," 423; see also Clary-Lemon, *Planting the Anthropocene*, 9–11.

67. Weisman, *World Without Us*, 5.

68. Watts, "We Have 12 Years."

69. Bitzer, "Rhetorical Situation," 7.

70. Green et al., "Facts."

71. Extinction Rebellion, "Our Demands," accessed January 17, 2021, https://extinctionrebellion.uk/the-truth/demands.

72. Langer, "Interview with Gail Bradbrook."

73. Delicath and DeLuca, "Image Events."

74. Extinction Rebellion, "The Blood of Our Children," YouTube video, 2:19, March 13, 2019, https://www.youtube.com/watch?v=r89CvYYKdxw.

75. As such, it brings together what Lind describes as the "archive" and the "advocacy" messaging of discourse about anthropogenic mass extinction ("*Unbearable Loss*," v.).

76. Extinction Rebellion NYC (@XR_NYC), "#ExtinctionRebellion New York City," Twitter, April 14, 2019, 10:59 a.m., https://twitter.com/XR_NYC/status/1117457331573415937.

77. Van Dooren, *Flight Ways*, quoted in Haraway, *Staying with the Trouble*, 39.

78. Extinction Rebellion, "BREAKING: Extinction Rebellion Holds Funeral Service on Parl Sq, Blocks Square," November 24, 2018, https://rebellion.earth/2018/11/24/breaking-extinction-rebellion-funeral-service-on-parl-sq-blocks-square.

79. Butler, *Precarious Life*; Butler, *Frames of War*. For the use of this idea in an environmental concept, see Mortimer-Sandilands and Erickson, *Queer Ecologies*; van Dooren, *Flight Ways*; Heise, *Imagining Extinction*; Haraway, *Staying with the Trouble*; Barnett, "Thinking Ecologically"; and Barnett, "Naming."

80. Haraway, *Staying with the Trouble*, 39; Barnett, "Naming," 297.

81. Butler, "Violence, Mourning, Politics," 17.

82. Wretched of the Earth, "Open Letter."

83. Whyte, "Our Ancestors' Dystopia Now," 208. Whyte is describing here a point by the environmental advocate Lee Sprague (Potawatomi / Odawa, Little River Band of Ottawa Indians). For more on this idea, see also Dillon, *Walking the Clouds*; Whyte, "Indigenous Science (Fiction) for the Anthropocene"; and Davis and Todd, "On the Importance."

84. Vizenor, *Manifest Manners*.

85. Davis and Todd, "On the Importance."

86. Pezzullo and Cox, *Environmental Communication*, 271.

87. Whyte, "Dakota Access Pipeline," 167. See also Farrell et al., "Effects of Land Dispossession."

88. Though the declaration was close in time to the XR action, the exigency for it was the Vuntut Gwitchin's annual Caribou days.

89. The declaration can be found on the Vuntut Gwitchin's website: http://www.vgfn.ca/pdf/CC%202019%20Declaration.pdf.

90. NOAA Arctic Program, "Arctic Report Card: Update for 2018," accessed January 19, 2021, https://arctic.noaa.gov/Report-Card/Report-Card-2018.

91. James, "We Are the Ones," 262. "Gwitchin" is also spelled "Gwich'in" elsewhere, especially when referring to the larger nation of which the Vuntut Gwitchin are a part, as James does here. In this chapter, I have opted to use the spelling used on the declaration and the Vuntut Gwitchin's website.

92. Extinction Rebellion US, "We Demand," accessed June 8, 2022, https://extinctionrebellion.us/demands.

93. Quoted in Dembicki, "Debate over Racism."

94. What happened in XR followed a familiar pattern by white activists when challenged to account for the issues of inclusion, injustice, and racism that have plagued the environmental movement from its inception. For more, see Pezzullo and Sandler, "Introduction."

95. Quoted in Dembicki, "Debate over Racism."

96. Extinction Rebellion America, "The XR America Seven Commitments," accessed June 8, 2022, https://drive.google.com/file/d/1htF_asRzrLIynLMMnU2f3v1eb77zvxKw/view.

97. Mbembe, *Out of the Dark Night*, 192.

Chapter 4

1. McKay et al., "Search for Past Life," 929.

2. Quoted in Wilford, "Clues in Meteorite."

3. William Clinton, "Statement Regarding Mars Meteorite Discovery," speech, August 7, 1996, http://www2.jpl.nasa.gov/snc/clinton.html.

4. For an overview of the scientific controversies regarding life detection in ALH84001 and the Viking missions, see Markley, *Dying Planet*, 303–54.

5. Quoted in Twilley, "Meet the Martians."

6. Quoted in Daly and Frodeman, "Separated at Birth," 146.

7. Sagan, *Cosmos*, 138.

8. Remarkably, the first genetic map of an archaeon in 1996 revealed that archaea are more genetically similar to humans than they are to bacteria, which suggests that we both descend from the microorganisms from which eukaryotes evolved. Bult et al., "Complete Genome Sequence." Enabled by technological developments in imaging and genomics, as well as biomedical interest in human microbiomes, microbiology has recently—and with breathtaking speed—moved from side to center stage in the life sciences, rumbling paradigms across a number of fields and reshaping the "tree of life" such that not just humans but *animalia*

constitutes only a thin branch. See, for example, the visual system proposed in Woese, Kandler, and Wheelis, "Towards a Natural System."

9. Johnson et al., "Ancient Bacteria."

10. Quoted in Hird, *Origin of Sociable Life*, 21, 142.

11. Consider, for example, a 1976 article by the British physician Bernard Dixon, who wondered whether there was a case to be made for conservation of smallpox. Eradicating smallpox, he argued, was "the first time in history when man has been able to obliterate—for all time and by conscious rational choice—a particular form of life" ("Smallpox"). Whether the smallpox virus *is* a form of life is, of course, subject to debate. Dixon argued that smallpox ought to be conserved because of its value to biomedical research, not because it had intrinsic value itself.

12. Gilbert, Sapp, and Tauber, "Symbiotic View."

13. Cockell, "Environmental Ethics and Size," 24.

14. Farrell, "Weight of Rhetoric," 472.

15. Hawhee, *Rhetoric in Tooth*, 56.

16. Farrell, "Sizing Things Up," 11.

17. Farrell, "Weight of Rhetoric," 484.

18. Smith, "Curious Case," 195–96. Credit for coining the term "astrobioethicists" goes to Brandon Keim in a brief blog post on *Wired* in 2008 ("Be a Parent"). While there are some tentative moves from European geoscientists to gather together under the name, as of the writing of this book, there is not yet a discipline of astrobioethics.

19. Squier, *Liminal Lives*, 264.

20. Farrell suggests that place is definitive of magnitude, though he doesn't say much about it ("Weight of Rhetoric," 473).

21. For more on deixis and its relationship to place, see Prasch, "Toward a Rhetorical Theory of Deixis."

22. Quoted in Faust, "AL84001+10."

23. Bada, "Field with a Life."

24. National Library of Medicine, "Launching a New Science."

25. Lederberg, "Sputnik + 30," 217.

26. Ibid. In 1958, Lederberg explained that he had occupied himself with "lunar, or cosmic microbiology" since December 1957, but for the "past several years," he said, "I wondered whether Arhenius [*sic*] should be discredited in discussions of biopoeisis + the question of contamination was my first reaction to speculation on a lunar probe since Sputnik 1." Untitled memo, February 24, 1958, Joshua Lederberg Papers, National Library of Medicine Profiles in Science, Bethesda, MD, http://www.profiles.nlm.nih.gov/BB.

27. Bradley, *Joshua Lederberg*, 11.

28. Lederberg and Cowie, "Moondust"; Schmeck, "Rockets' Effect on Moon." In response to a question about how the exobiology community formed, Lederberg joked, "well, it was mobilized by the daily papers, above all." Lederberg, interview by Steven J. Dick, November 12, 1992, transcript, Lederberg Papers.

29. Lederberg, "Exobiology: Approaches," 393.

30. Strick, "Creating a Cosmic Discipline," 139.

31. Rosen, *Life Itself*, 6.

32. "Matter and energy aspire to be universal concepts; life in contemporary science still means terrestrial life," Lederberg wrote. National Academy of Sciences, National Research Council, Space Science Board, "WESTEX Summary Report" Draft, October 16, 1959, Lederberg Papers.

33. Lederberg, "Exobiology: Approaches," 394.

34. Lederberg, "Signs of Life."

35. See Mody, "A Little Dirt."

36. Redd, "Planetary Protection." Avoiding scientific contamination remains one of the prime directives of NASA's Office of Planetary Protection, which is a part of the Office of Safety and Mission Assurance.

37. Many of the scientists who became involved in exobiology were already working on the origin of life question, one of many grails occupying biologists in the 1950s. Watson and Crick's first paper on the structure of DNA hit in 1953. In 1957, the Russian scientist Aleksandr Oparin, author of the 1938 book *The Origins of Life*, convened the first international gathering on the topic. Stanley Miller and Harold Urey's paper on creating the "spark of life" in a test tube was published in 1959. For more, see Strick, "Creating a Cosmic Discipline."

38. Temple, "Prehistory of Panspermia." Only fragments of Anaxagoras's work survive, and his theory of panspermia has been reconstructed through paraphrases from others, such as Theophrastus, who wrote that "Anaxagoras says that the air contains the seeds of all things, and that these, carried down by the rain, produce the plants" (*Enquiry into Plants*, 163).

39. Bound as the history of Western science is with colonialism, it is no surprise that when contemporary scientists started thinking seriously about panspermia two millennia later, it is Anaxagoras rather than the ancient Egyptians or Indians who are given credit for this idea, and it is the Greek word that scientists continue to use.

40. For an overview of the early scientific interest in panspermia, see Demets, "Darwin's Contribution." Panspermia and theories like it continue to attract attention from contemporary scientists. Consider a wonderfully weird article recently published in *Progress in Biophysics and Molecular Biology* that points to the abrupt appearance of the nervous system of the octopus in the evolutionary record as one potential piece of evidence that life did not originate on Earth. The octopus has long confounded evolutionary biologists, as some of the striking features of the species cannot be traced to the nautilus, from which it supposedly descends. The authors write, "One plausible explanation, in our view, is that *the new genes are likely new extraterrestrial imports to Earth*—most plausibly as an already coherent group of functioning genes within (say) cryopreserved and matrix protected fertilised octopus eggs," which would "be a parsimonious cosmic explanation for the octopus' sudden emergence on Earth [about] 270 million years ago" (Steele et al., "Cause of Cambrian Explosion," 4; my emphasis).

41. Lederberg, "Exobiology: Approaches," 396.

42. Wolfe, "Germs in Space," 189. While Lederberg uses panspermia as an example in "Exobiology," elsewhere he cautioned that while not "*so* implausible that they should be totally ignored," he found that "plead[ing] our basic ignorance" would be more rhetorically effective to policy makers than proposing the particular mechanism of panspermia as a scientific justification (ibid.).

43. Lederberg, "Exobiology: Approaches," 396. Lederberg thought the interdisciplinary nature of exobiology was of such paramount importance that he strongly objected to the formation of an exobiology journal, which he felt would "isolate the field from the badly needed critical judgments" of the scientific community writ large (Lederberg, "Exobiology," 1126).

44. Lederberg, "Exobiology: Approaches," 400.

45. Joshua Lederberg to Vladimir D. Timakov, May 16, 1958, Lederberg Papers.

46. Rummel and Billings, "Issues in Planetary Protection," 50; Derbyshire, "Resumé," 11.

47. Article IX from United Nations Resolution 2222, Treaty on Principles Governing the Activities of States in the Exploration and Use of Outer Space, Including the Moon and Other Celestial Bodies (UN Doc. A/RES/2222), signed January 27, 1967, https://www.unoosa.org/oosa/en/ourwork/spacelaw/treaties/outerspacetreaty.html.

48. COSPAR (Committee on Space Research), "Panel on Planetary Protection," January 13, 2021, https://cosparhq.cnes.fr/scientific-structure/panels/panel-on-planetary-protection-ppp.

49. Daly and Frodeman, "Separated at Birth," 141.

50. Kiminek et al., "COSPAR's Planetary Protection Policy."

51. Douglas, *Purity and Danger*.

52. In NASA materials, what I am calling "scientific contamination" is under the umbrella of forward contamination. I have distinguished scientific and forward contamination in this chapter so as to highlight the different metrics of value by which interplanetary contamination is understood to be a problem.

53. Wells, *War of the Worlds*, 282.

54. Ibid., 240–41 (my emphasis). Note that this passage is not in the original *Pearson's* serial of the novel.

55. Ibid., 207.

56. Ibid., 283.

57. Leopold, *Sand County Almanac*, 192.

58. The concerns about contamination on the Apollo missions only extended to protecting the Earth from potential moon microbes—the moon seems to be of little concern. Every mission to the moon left trash on its surface, including bags of human waste that very well may harbor still-living microbes. Scientists are now confident that there is no (native) life on the moon, and any missions there would be classified Category 1—"not of direct interest for understanding the process of chemical evolution or the origin of life"—and thus not subject to sterilization protocols that would avoid scientific contamination (Garber, "Trash We've Left Behind").

59. Quoted in Stone, *Chasing the Moon*.

60. See Van Vechten Trumbull to Joshua Lederberg, July 29, 1969; Joshua Lederberg to Harry Schwartz, May 29, 1969; and Dennis W. Watson to Asger F. Langlykke, July 31, 1969, all in Lederberg Papers.

61. "Is the Earth Safe."

62. Joshua Lederberg, "Lunar Quarantine," letter to the *New York Times*, July 13, 1969, Lederberg Papers.

63. Robert J. Ferl and Anna-Lisa Paul also make the point that science fiction had primed the American public to be suspicious of the dangers of extraterrestrial life in the 1960s ("Lunar Plant Biology," 274). See also Meltzer, *When Biospheres Collide*, 215.

64. Hsu, "Myth."

65. See Jordan, "Kennedy's Romantic Moon." Jordan argues that to overcome concerns about the costs and risks of lunar exploration, President Kennedy used romantic, transcendent images to involve the American public as partners in the mission.

66. Raymont, "Publishers Hitching Star."

67. His concern was warranted. Crichton's protagonist, Jeremy Stone, is also a professor of bacteriology at Stanford and described as "thin and balding." Additionally, in the novel, Stone was awarded the Nobel Prize for his work on bacteria at age thirty-six, after which he turned his attention to the issue interplanetary contamination, sending articles titled "Sterilization of Spacecraft" to *Science* and *Nature*. He convenes a group of like-minded, interdisciplinary scientists to discuss the issues involved in "the contamination problem." In a letter to the Universal Studios producer Robert Wise—and cc'ed to Carl Sagan—Lederberg wrote that he hoped Wise would "take reasonable and prudent precautions to minimize the possibility of public confusion between [himself] and any of the characters in the novel." Joshua Lederberg to Robert Wise Productions, June 1969, Lederberg Papers.

68. Crichton, *Andromeda Strain*, 278.

69. Crichton, "The Andromeda Strain," Official Site of Michael Crichton, accessed January 13, 2020, http://www.michaelcrichton.com/the-andromeda-strain.

70. Crichton, *Andromeda Strain*, 3.

71. Schott, review of *Andromeda Strain*, BR4.

72. Quoted in Meltzer, *When Biospheres Collide*, 74.

73. Sullivan, "Guarding Against Moon Bugs," E8.

74. Meltzer, *When Biospheres Collide*, 215.

75. Wald, *Contagious*, 32.

76. Ibid., 58.

77. Ibid., 51. See also Chávez, *Borders of AIDS*, 19–40.

78. Sullivan, "Guarding Against Moon Bugs," E8.

79. NASA, "Planetary Protection."

80. Rummel, "Planetary Exploration," 2130.

81. International Committee Against Mars Sample Return, "Charter," accessed August 7, 2018, http://www.icamsr.org/index.html.

82. Quoted on http://icamsr.org, accessed September 12, 2019. This quotation is taken from a personal communication between Woese and Barry DiGregorio, the founder of ICAMSR.

83. Rummel, "Planetary Exploration."

84. Redd, "Planetary Protection."

85. Meltzer, *When Biospheres Collide*, xv, 216–18.

86. Wells, *War of the Worlds*, 7. For further commentary on this passage, see Brantlinger, *Taming Cannibals*, 185; for more on the imperial themes in Wells, see Seed, "The Course of Empire."

87. Wells, *War of the Worlds*, 196. It's worth noting that the sentence "Should we conquer?"—which is found in the *Pearson's* serial as well as the first edition of the novel—was edited out of many subsequent editions of the book, such as the Signet Classics edition currently in print. J. Jesse Ramírez makes a compelling case that anti-imperialist elements of *War of the Worlds* were deliberately omitted in many American adaptations of Welles's text ("From Woking").

88. Wolfe, "Germs in Space," 188.

89. Lederberg, "Exobiology: Approaches," 398.

90. In addition to inherent values, Christine Oravec names recreation and scenic values as motivational forces in the preservationist movement in the US ("John Muir," 245).

91. A "moral patient" is "a legitimate object of moral concern: that is, roughly, [it is] an entity that has interests that should be taken into considerations when decisions are made concerning it or which otherwise impact on it" (Rowlands, *Can Animals Be Moral?*, 72).

92. NASA's Office of Planetary Protection, for example, names its directive to "carefully control forward contamination of other worlds by terrestrial organisms and organic materials carried by spacecraft in order to guarantee the integrity of the search and study of extraterrestrial life, if it exists." NASA, "Planetary Protection," accessed January 16, 2021, https://sma.nasa.gov/sma-disciplines/planetary-protection.

93. Sagan, *Cosmos*, 130. Despite his concern for potential Martian life, Sagan was not a strict preservationist, and if no evidence of life was to be found, he was very much in favor of exploring Mars. Shortly before his death in 1996, in fact, Sagan recorded a message to future explorers that was included on board the 2008 Phoenix Mars Lander: "Maybe we're on Mars because of the magnificent science that can be done there—the gates of the wonder world are opening in our time. Maybe we're on Mars because we have to be, because there's a deep

nomadic impulse built into us by the evolutionary process—we come, after all, from hunter-gatherers, and for 99.9% of our tenure on Earth we've been wanderers. And the next place to wander to is Mars. But whatever the reason you're on Mars is, I'm glad you're there. And I wish I was with you" (National Public Radio, "Science Fiction"). As these remarks make clear, Sagan's preservationism was not about Mars, per se, but *life on Mars*.

94. McKay, "Biologically Reversible Exploration."

95. Randolph, Race, and McKay, "Reconsidering the Theological," 4. See also McKay, "Planetary Ecosynthesis."

96. Lupisella, "Rights of Martians," 94.

97. Randolph and McKay, "Protecting and Expanding," 30.

98. Ibid., 31 (my emphasis).

99. Of course, even an abiotic Mars has instrumental value, as well as economic value. The going rate for Martian rocks (in the form of meteorites) is between $11,000 and $22,500 an ounce—ten times the price of gold. Tarantola, "New Mars Meteorite."

100. P. Taylor, *Respect for Nature*, 61. While Taylor believed that all living things were moral patients, he did not believe that all living things were subject to equal moral consideration: "a human's good, just because it is a human's, would always outweigh the good of an animal or plant" (133).

101. Grinspoon, "Is Mars Ours?"

102. McKay, "Does Mars Have Rights?," 194.

103. Zubrin, "Case for Terraforming," 179–80, quoted in Daly and Frodeman, "Separated at Birth," 146.

104. Zubrin, *Case for Mars*, 267.

105. Daly and Frodeman, "Separated at Birth," 147, quoting Marshall, "Ethics and the Extraterrestrial," 234.

106. Daly and Frodeman, "Separated at Birth," 148.

107. Robinson, *Green Mars*, 1.

108. Robinson, *Red Mars*, 175 (my emphasis).

109. Robinson, *Green Mars*, 30; Robinson, *Red Mars*, 174.

110. Robinson, *Red Mars*, 85.

111. Ibid., 174.

112. Ibid., 145.

113. Ibid., 229.

114. Here I am tweaking a phrase from the Mayan priest Nicolas Lucas Ticum, who, in a 2009 presentation to the UN Permanent Forum on Indigenous Issues, stated, "The Earth does not belong to human beings; human beings belong to the Earth" (United Nations, "'The Earth Does Not Belong to Human Beings; Human Beings Belong to the Earth,' Permanent Forum Hears as It Takes Up Issues of Climate Change, Land Tenure," Meetings Coverage, May 27, 2009, https://www.un.org/press/en/2009/hr4988.doc.htm). This phrase is also found in a well-known speech that may or may not have been composed by Chief Seattle. For more on that text, and a discussion of its authenticity and authority, see Burkhart, *Indigenizing Philosophy*.

115. Messeri, *Placing Outer Space*, 196.

116. Jennifer Clary-Lemon writes that "there is no 'I am' that stands outside of relation" (*Planting the Anthropocene*, 64).

117. The markings that Lowell called "canals" had first been described in 1877 by the Italian astronomer Giovanni Schiaparelli as *canali* (which in Italian means "channels," not "canals"). Schiaparelli later walked back this claim. For a detailed history of Lowell and the canal controversy, see Markley, *Dying Planet*, 61–114.

118. Sheehan and Dobbins, "Spokes of Venus," 60–62.

119. J. DiCaglio, *Scale Theory*, 239.

120. Kanas and Manzey, *Space Psychology and Psychiatry*. See also Hersch, "Space Madness."

121. Robinson, *Red Mars*, 224.

122. Pezzullo highlights Aristotle's argument in *De Anima* that there is "something of the marvelous" in "all things of nature" ("Unearthing the Marvelous," 25).

123. Markley writes that Mars has long been a "screen" on which humans project images of ourselves and the meaning of Earth (*Dying Planet*, 2).

Conclusion

1. Mann, *Magic Mountain*, 326.

2. Ibid., 328.

3. Achille Mbembe, remarks to the Alien Earth Mellon-Borghesi Workshop, University of Wisconsin–Madison, May 5, 2021.

4. Stuart Murray urges us to consider how what Foucault called "thanatopolitics" is not just an aspect or outcome of biopolitics but its "constitutive onto-logic" ("Affirming the Human?," 492). Jasbir Puar echoes this line of thinking, arguing that biopolitics "makes its presence known at the limits and through the excess of [necropolitics]; the former masks the multiplicity of its relationships to death and killing in order to enable the proliferation of the latter" (*Terrorist Assemblages*, 35). Puar argues that it is imperative to hold a "bio-necro tension" in mind (35), which Rowland suggests "demands attunement to the deathly imperative in every life-building project" (*Zoetropes*, 23–24).

5. Murphy, *Economization of Life*, 141.

6. Looking back through Mbembe's body of work, I was surprised to find how frequently he invoked life in this way. See, for example, *Necropolitics*, 3, 180–81.

7. Rose, *Politics of Life Itself*, 3. In *Out of the Dark Night*, Mbembe calls for a reimagination of the "common that includes nonhumans, which is the proper name for a democracy" (41).

8. Mbembe, "Universal Right," S60.

9. Ibid., S62.

10. Ibid. The "planetary life" phrase is from Mbembe, *Out of the Dark Night*, in a passage where he notes that the age of the human as marked by agency has come to an end, and thus "the times are propitious for a return to 'big questions' and 'deep history'—'big questions' concerning the relation of human life to planetary life, in a context of geological recasting of historical time" (88).

11. Mbembe, *Critique of Black Reason*, 183.

12. Mbembe, "Universal Right," S60.

13. I am reminded of a striking line in Kevin Browne's essay "No Words" about the limits of language in the face of Black death: "*Breathing*, for those who cannot breathe, is a right" (17). Mbembe calls for breathing as a universal right, but Browne reminds us that rights take their most coherent, and most urgent, shape when foreclosed.

14. See Nealson et al. "Breathing Metals."

15. Margulis and Sagan, *What Is Life?*, 23.

16. S. DiCaglio, "Breathing in a Pandemic," 376.

17. Mbembe, *Out of the Dark Night*, 88.

Bibliography

Abbey, Edward. "Immigration and Liberal Taboos." In *One Life at a Time, Please*, 41–44. New York: Henry Holt, 1988.

Adams, Katherine. "At the Table with Arendt: Toward a Self-Interested Practice of Coalition Discourse." *Hypatia* 17, no. 1 (2002): 1–33.

Ahmed, Sara. *The Cultural Politics of Emotion*. Edinburgh: Edinburgh University Press, 2004.

American Horse, Iyuskin. "We Are Protectors, Not Protestors." *The Guardian*, August 18, 2016. https://www.theguardian.com/us-news/2016/aug/18/north-dakota-pipeline-activists-bakken-oil-fields.

Anker, Peder. "Deep Ecology in Bucharest." *The Trumpeter* 24, no. 1 (2008): 56–58.

Arendt, Hannah. *Between Past and Future: Eight Exercises in Political Thought*. New York: Penguin, 1993.

———. *The Human Condition*. 2nd ed. Chicago: University of Chicago Press, 2013.

Aristotle. *De Anima*. Translated by Christopher Shields. Oxford: Oxford University Press, 2009.

Arrhenius, Svante. *Worlds in the Making—The Evolution of the Universe*. Translated by H. Borns. New York: Harper, 1908.

Arvin, Maile. "Analytics of Indigeneity." In *Native Studies Keywords*, edited by Stephanie Nohelani Teves, Andrea Smith, and Michelle Raheja, 119–29. Tucson: University of Arizona Press, 2015.

Bada, Jeffrey L. "A Field with a Life of Its Own." *Science* 307, no. 5706 (2005): 46.

Barkun, Michael. "Divided Apocalypse: Thinking About the End in Contemporary America." *Soundings: An Interdisciplinary Journal* 66, no. 3 (1983): 257–80.

Barnett, Joshua Trey. "Naming, Mourning, and the Work of Earthly Coexistence." *Environmental Communication* 13, no. 3 (2019): 287–99.

———. "Thinking Ecologically with Judith Butler." *Culture, Theory, and Critique* 59, no. 1 (2018): 20–39.

Barnosky, Anthony D., Nicholas Matzke, Susumu Tomiya, Guinevere O. U. Wogan, Brian Swartz, Tiago B. Quental, Charles Marshall, Jenny L. McGuire, Emily L. Lindsey, Kaitlin C. Maguire, Ben Mersey, and Elizabeth A. Ferrer. "Has the Earth's Sixth Mass Extinction Already Arrived?" *Nature* 471 (2011): 51–57.

Bedau, Mark, and Carol Cleland. Introduction to *The Nature of Life: Classical and Contemporary Perspectives from Philosophy and Science*, edited by Bedau and Cleland, xix–xxii. Cambridge: Cambridge University Press, 2010.

Benhabib, Seyla. *The Reluctant Modernism of Hannah Arendt*. Oxford: Rowman & Littlefield, 2000.

Bennett, Jane. *Vibrant Matter: A Political Ecology of Things*. Durham: Duke University Press, 2010.

Benton, E. "Vitalism in Nineteenth Century Thought." *Studies in the History and Philosophy of Science* 5 (1974): 17–48.

Bernasconi, Robert. "The Double Face of the Political and the Social: Hannah Arendt and America's Racial Divisions." *Research in Phenomenology* 26 (1996): 3–24.

Biesecker, Barbara. *Addressing Postmodernity: Kenneth Burke, Rhetoric, and a Theory of Social Change*. Tuscaloosa: University of Alabama Press, 2000.

Bitzer, Lloyd F. "The Rhetorical Situation." *Philosophy and Rhetoric* 1, no. 1 (1968): 1–14.

Boes, Tobias. "Beyond Whole Earth: Planetary Mediation and the Anthropocene." *Environmental Humanities* 5, no. 1 (2014): 155–70.

Bonilla-Silva, Eduardo. *Racism Without Racists: Color-Blind Racism and the Persistence of Racial Inequality in the United States*. New York: Rowman and Littlefield, 2006.

Bookchin, Murray. "The Crisis in the Ecology Movement." *Z Magazine* 6 (1988). Accessed June 8, 2022. https://theanarchistlibrary.org/library/murray-bookchin-the-crisis-in-the-ecology-movement.

———. "Social Ecology Versus Deep Ecology." *Green Perspectives: Newsletter of the Green Program Project* 4–5 (1987). Accessed June 8, 2002. https://theanarchistlibrary.org/library/murray-bookchin-social-ecology-versus-deep-ecology-a-challenge-for-the-ecology-movement.

Bookchin, Murray, and Dave Foreman. *Defending the Earth*. Montreal: Black Rose Books, 1991.

Bradley, S. Gaylen. *Joshua Lederberg, 1925–2008: A Biographical Memoir*. Washington, DC: National Academy of Sciences, 2009.

Brand, Stewart. *Space Colonies*. Harmondsworth, UK: Penguin, 1977.

Brantlinger, Patrick. *Taming Cannibals: Race and the Victorians*. Ithaca: Cornell University Press, 2011.

Browne, Kevin Adonis. "No Words." *Brick* 106 (Winter 2021): 8–17.

Bryant, Levi. *Democracy of Objects*. London: Open Humanities Press, 2011.

Buell, Frederick. *From Apocalypse to Way of Life: Environmental Crisis in the American Century*. New York: Routledge, 2003.

Bult, Carol J., et al. "Complete Genome Sequence of the Methanogenic Archaeon, Methanococcus Jannaschii." *Science* 273, no. 5278 (1996): 1058–73.

Burke, Edmund. *A Philosophical Enquiry into the Origin of Our Ideas of the Sublime and Beautiful*. New York: Columbia University Press, 1958. First published 1757.

Burke, Kenneth. *Grammar of Motives*. Berkeley: University of California Press, 1969. First published 1945.

———. *Language as Symbolic Action*. Berkeley: University of California Press, 1966.

———. "Methodological Repression and/or Strategies of Containment." *Critical Inquiry* 5, no. 2 (1978): 401–16.

———. *Rhetoric of Motives*. Berkeley: University of California Press, 1969. First published 1950.

Burkhart, Brian Yazzie. "Be as Strong as the Land That Made You: An Indigenous Philosophy of Well-Being Through the Land." *Science, Religion and Culture* 6, no. 1 (2019): 26–33.

———. *Indigenizing Philosophy Through the Land: A Trickster Methodology for Decolonizing Environmental Ethics and Indigenous Futures*. East Lansing: Michigan State University Press, 2019.

Butler, Judith. *Frames of War: When Is Life Grievable?* London: Verso, 2016.

———. *Precarious Life: The Powers of Mourning and Violence*. London: Verso, 2006.

———. "Violence, Mourning, Politics." *Studies in Gender and Sexuality* 4, no. 1 (2003): 9–37.

Cagle, Susie. "'Bees, Not Refugees': The Environmentalist Roots of Anti-Immigrant Bigotry." *The Guardian*, August 16, 2019. https://www.theguardian.com/environment/2019/aug/15/anti.

Callicott, J. B. "On the Intrinsic Value of Non-human Species." In *The Preservation of Species*, edited by Bryan G. Norton, 138–72. Princeton: Princeton University Press, 1986.

Canguilhem, Georges. *Knowledge of Life*. New York: Fordham University Press, 2009.

Castro-Gómez, Santiago. *La hybris del punto cero: Ciencia, raza e ilustración en la Nueva Granada (1750–1816)*. Bogotá: Editorial Pontificia Universidad Javeriana, 2005.

Ceballos, Gerardo, Paul R. Ehrlich, Anthony D. Barnosky, Andrés García, Robert M. Pringle, and Todd M. Palmer. "Accelerated Modern Human-Induced Species Losses: Entering the Sixth Mass Extinction." *Science Advances* 1, no. 5 (2015): 1–5.

Ceballos, Gerardo, Paul R. Ehrlich, and Rodolfo Dirzo. "Biological Annihilation via the Ongoing Sixth Mass Extinction Signaled by Vertebrate Population Losses and Declines." *Proceedings of the National Academy of Sciences* 114, no. 30 (2017): E6089–E6096.

Chakrabarty, Dipesh. "The Climate of History: Four Theses." *Critical Inquiry* 35, no. 2 (2009): 197–222.

Charland, Maurice. "Constitutive Rhetoric: The Case of the *Peuple Québécois*." *Quarterly Journal of Speech* 73, no. 2 (1987): 133–50.

Chávez, Karma R. *The Borders of AIDS: Race, Quarantine, and Resistance*. Seattle: University of Washington Press, 2021.

———. *Queer Migration Politics: Activist Rhetoric and Coalitional Possibilities*. Urbana: University of Illinois Press, 2013.

Chen, Mel Y. *Animacies: Biopolitics, Racial Mattering, and Queer Affect*. Durham: Duke University Press, 2012.

Chen, S. *A Documentary on Shenzhou-9*. Hunan, China: Science and Technology Press of Hunan, 2012.

Clary-Lemon, Jennifer. *Planting the Anthropocene: Rhetorics of Natureculture*. Logan: Utah State University Press, 2019.

Coates, Peter. *American Perceptions of Immigrant and Invasive Species: Strangers on the Land*. Berkeley: University of California Press, 2007.

Cockell, Charles S. "Environmental Ethics and Size." *Ethics and the Environment* 13, no. 1 (2008): 23–39.

Cosgrove, Denis. "Contested Global Visions: One-World, Whole-Earth, and the Apollo Space Photographs." *Annals of American Geographers* 84, no. 2 (June 1994): 270–94.

Cox, J. Robert. "The Die Is Cast: Topical and Ontological Dimensions of the Locus of the Irreparable." *Quarterly Journal of Speech* 68, no. 3 (1982): 227–39.

Crichton, Michael. *The Andromeda Strain*. New York: Vintage Books, 1969.

Dalrymple, Amy. "Pipeline Route Plan First Called for Crossing North of Bismark." *Bismarck Tribune*, November 1, 2016.

Daly, Erin Moore, and Robert Frodeman. "Separated at Birth, Signs of Rapprochement: Environmental Ethics and Space Exploration." *Ethics and the Environment* 13, no. 1 (2008): 135–51.

Davis, Diane. "Burke and Freud on Who You Are." *Rhetoric Society Quarterly* 38, no. 2 (2008): 123–47.

———. *Inessential Solidarity: Rhetoric and Foreigner Relations*. Pittsburgh: University of Pittsburgh Press, 2010.

Davis, Heather, and Zoe Todd. "On the Importance of a Date, or, Decolonizing the Anthropocene." *ACME: An International Journal for Critical Geographies* 16, no. 4 (2017): 761–80.

Delicath, John W., and Kevin Michael DeLuca. "Image Events, the Public Sphere, and Argumentative Practice: The Case of Radical Environmental Groups." *Argumentation* 17 (2003): 315–33.

Deloria, Vine, Jr. "American Indian Metaphysics." In *Power and Place: Indian Education in America*, edited by Vine Deloria Jr. and Daniel R. Wildcat, 1–6. Golden, CO: Fulcrum, 2001.

Dembicki, Geoff. "A Debate over Racism Has Split One of the World's Most Famous Climate Groups." Vice, April 28, 2020. https://www.vice.com/en/article/jgey8k/a-debate-over-racism-has-split-one-of-the-worlds-most-famous-climate-groups.

Demets, Rene. "Darwin's Contribution to the Development of the Panspermia Theory." *Astrobiology* 12, no. 10 (2012): 946–50.

Derbyshire, G. A. "Resumé of Some Earlier Extraterrestrial Contamination Activities." In *A Review of Space Research*, edited by Space Science Board, 10.11–10.13. Washington, DC: National Academy of Sciences, 1962.

Derrida, Jacques. *The Animal That Therefore I Am*. New York: Fordham University Press, 2008.

———. *Aporias: Dying—Awaiting (One Another at) the "Limits of Truth."* Stanford: Stanford University Press, 1993.

de Vos, Jurriaan M., Lucas N. Joppa, John L. Gittleman, Patrick R. Stephens, and Stuart L. Pimm. "Estimating the Normal Background Rate of Species Extinction." *Conservation Biology* 29, no. 2 (2015): 452–62.

DiCaglio, Joshua. *Scale Theory: A Nondisciplinary Inquiry*. Minneapolis: University of Minnesota Press, 2021.

DiCaglio, Joshua, Kathryn M. Barlow, and Joseph S. Johnson. "Rhetorical Recommendations Built on Ecological Experience: A Reassessment of the Challenge of Environmental Communication." *Environmental Communication* 12, no. 4 (2018): 438–50.

DiCaglio, Sara. "Breathing in a Pandemic: Covid-19's Atmospheric Erasures." *Configurations* 29, no. 4 (2021): 375–87.

Diehm, Christian. "Ethics and Natural History: Levinas and Other-than-Human Animals." *Environmental Philosophy* 3, no. 2 (2006): 34–43.

———. "Identification with Nature: What It Is and Why It Matters." *Ethics and the Environment* 12, no. 2 (2007): 1–22.

Dillon, Grace L., ed. *Walking the Clouds: An Anthology of Indigenous Science Fiction*. Tucson: University of Arizona Press, 2012.

Dixon, Bernard. "Smallpox—Imminent Extinction and an Unresolved Dilemma." *New Scientist* 69, no. 989 (1976): 430–32.

Douglas, Mary. *Purity and Danger: An Analysis of Concepts of Pollution and Taboo*. London: Routledge, 2013.

Doyle, Richard. *On Beyond Living: The Rhetorical Transformations of the Life Sciences*. Stanford: Stanford University Press, 1997.

———. *Wetwares: Experiments in Postvital Living*. Minneapolis: University of Minnesota Press, 2003.

Druschke, Caroline Gottschalk, and Candance Rai. "Making Worlds with Cyborg Fish." In *Tracing Rhetoric and Material Life: Ecological Approaches*, edited by Bridie McGreavy, Justine Wells, George F. McHendry Jr., and Samantha Senda-Cook, 197–222. Cham, Switzerland: Springer, 2017.

Durham, Weldon B. "Kenneth Burke's Concept of Substance." *Quarterly Journal of Speech* 66, no. 4 (1980): 351–64.

Estes, Nick, and Jaskiran Dhillon, eds. *Standing with Standing Rock: Voices from the #NoDAPL Movement*. Minneapolis: University of Minnesota Press, 2019.

Fanon, Frantz. *The Wretched of the Earth*. New York: Grove, 2007. First published 1961.

Farrell, Justin, Paul Berne Burow, Kathryn McConnell, Jude Bayham, Kyle Whyte, and Gal Koss. "Effects of Land Dispossession and Forced Migration on Indigenous Peoples in North America." *Science* 374, no. 6567 (2021): eabe4943.

Farrell, Thomas B. "Sizing Things Up: Colloquial Reflection as Practical Wisdom." *Argumentation* 12, no. 1 (1998): 1–14.

———. "The Weight of Rhetoric: Studies in Cultural Delirium." *Philosophy and Rhetoric* 41, no. 4 (2008): 467–87.

Faust, Jeff. "AL84001+10." *Space Review*, August 7, 2006. http://www.thespacereview.com/article/678/1.

Feinberg, Matthew, and Robb Willer. "Apocalypse Soon? Dire Messages Reduce Belief in Global Warming by Contradicting Just-World Beliefs." *Psychological Science* 22, no. 1 (2011): 34–38.

Ferl, Robert J., and Anna-Lisa Paul. "Lunar Plant Biology—A Review of the Apollo Era." *Astrobiology* 10, no. 3 (2010): 261–74.

Ferreira, Becky. "Seeing Earth from Space Is the Key to Saving Our Species from Itself." Vice, October 12, 2016. https://www.vice.com/en/article/bmvpxq/to-save-humanity-look-at-earth-from-space-overview-effect.

Foreman, Dave. *Confessions of an Eco-Warrior*. New York: Crown, 1991.

———. *Man Swarm and the Killing of Wildlife*. Devon, UK: Raven's Eye, 2011.

Foss, Sonja K., and Cindy L. Griffin. "A Feminist Perspective on Rhetorical Theory: Toward a Clarification of Boundaries." *Western Journal of Communication* 56 (1992): 330–49.

Foucault, Michel. *The Order of Things: An Archaeology of the Human Sciences*. New York: Pantheon, 1971.

Fox, Warwick. *Toward a Transpersonal Ecology: Developing New Foundations for Environmentalism*. Albany: SUNY Press, 1995.

French, Francis, and Colin Burgess. *In the Shadow of the Moon: A Challenging Journey to Tranquility, 1965–1969*. Lincoln: University of Nebraska Press, 2007.

Gagarin, Yuri. "Yuri Gagarin's First Speech About His Flight to Space." *The Atlantic*, April 12, 2011, https://www.theatlantic.com/technology/archive/2011/04/yuri-gagarins-first-speech-about-his-flight-into-space/237134. Speech originally given in Moscow, April 16, 1961.

Garber, Megan. "The Trash We've Left Behind on the Moon." *The Atlantic*, December 19, 2012. https://www.theatlantic.com/technology/archive/2012/12/the-trash-weve-left-on-the-moon/266465.

Garland-Thomson, Rosemarie. "Misfits: A Feminist Materialist Disability Concept." *Hypatia* 26, no. 3 (2011): 591–609.

Gehrke, Pat J. "The Ethical Importance of Being Human: God and Humanism in Levinas's Philosophy." *Philosophy Today* 50, no. 4 (2006): 428–36.

George, Allison. "Yuri Gagarin: 108 Minutes in Space." *New Scientist*, April 6, 2011. https://www.newscientist.com/article/mg21028075-600-yuri-gagarin-108-minutes-in-space.

Gelter, Hans. "Friluftsliv: The Scandinavian Philosophy of Outdoor Life." *Canadian Journal of Environmental Education (CJEE)* 5, no. 1 (2000): 77–92.

Gilbert, Scott F., Jan Sapp, and Alfred I. Tauber. "A Symbiotic View of Life: We Have Never Been Individuals." *Quarterly Review of Biology* 87, no. 4 (2012): 325–41.

Gilroy, Paul. *Postcolonial Melancholia*. New York: Columbia University Press, 2005.

Gines, Kathryn T. *Hannah Arendt and the Negro Question*. Bloomington: Indiana, University Press, 2014.

———. "Hannah Arendt, Liberalism, and Racism: Controversies Concerning Violence, Segregation, and Education." *Southern Journal of Philosophy* 47 (2009): 53–76.

Goedhart, Robert F. A. *The Neverending Dispute: Delimitation of Air Space and Outer Space*. Gif-sur-Yvette, France: Frontieres, 1996.

Goodpaster, Kenneth E. "On Being Morally Considerable." *Journal of Philosophy* 75, no. 6 (1978): 308–25.

Grant, David M. "Writing 'Wakan': The Lakota Pipe as Rhetorical Object." *College Composition and Communication* 69, no. 1 (2017): 61–86.

Green, Alison, et al. "Facts About our Ecological Crisis Are Incontrovertible. We Must Take Action." Letter to *The Guardian*, October 28, 2018. https:// www.theguardian.com/environment/2018/oct/26/facts-about-our-ecological-crisis-are-incontrovertible-we-must-take-action.

Grinspoon, David, "Is Mars Ours? The Logistics and Ethics of Colonizing the Red Planet." *Slate*, January 7, 2004. https://slate.com/technology/2004/01/the-logistics-and-ethics-of-colonizing-the-red-planet.html.

Gross, Alan. "Darwin's Diagram: Scientific Visions and Scientific Visuals." In *Ways of Seeing, Ways of Speaking: The Integration of Rhetoric and Vision in Constructing the Real*, edited by K. S. Fleckenstein, S. Hum, and L. T. Calendrillo, 52–80. New York: Parlor, 2007.

Gunn, Joshua, and David E. Beard. "On the Apocalyptic Sublime." *Southern Journal of Communication* 65, no. 4 (2000): 269–86.

Haldane, J. B. S. *What Is Life?* London: Alcuin, 1949.

Haraway, Donna. *Staying with the Trouble: Making Kin in the Cthulucene*. Durham: Duke University Press, 2016.

———. *When Species Meet*. Minneapolis: University of Minnesota Press, 2013.

Hawhee, Debra. *Moving Bodies: Kenneth Burke at the Edges of Language*. Columbia: University of South Carolina Press, 2009.

———. *Rhetoric in Tooth and Claw: Animals, Language, Sensation*. Chicago: University of Chicago Press, 2017.

Hawk, Byron. *A Counter-history of Composition: Toward Methodologies of Complexity*. Pittsburgh: University of Pittsburgh Press, 2007.

Heise, Ursula. *Imagining Extinction: The Cultural Meanings of Endangered Species*. Chicago: University of Chicago Press, 2016.

Henderson, Bob, and Nils Vikander. Introduction to *Nature First: Outdoor Life the Friluftsliv Way*, edited by Bob Henderson and Nils Vikander, 3–20. Toronto: Dundurn, 2007.

Hersch, Matthew H. "Space Madness: The Dreaded Disease That Never Was." *Endeavour* 36, no. 1 (2012): 32–40.

Hird, Myra. *The Origin of Sociable Life: Evolution after Science Studies*. New York: Palgrave, 2009.

Hitt, Christopher. "Toward an Ecological Sublime." *New Literary History* 30, no. 4 (1999): 603–23.

Hodges, Adam. "'Yes, We Can': The Social Life of a Political Slogan." In *Contemporary Critical Discourse Studies*, edited by Christopher Hart and Piotr Cap, 347–64. London: Bloomsbury, 2014.

Howe, Craig, and Tyler Young. "Mnisose." In *Standing with Standing Rock: Voices from the #NoDAPL Movement*, edited by Nick Estes and Jaskiran Dhillon, 56–70. Minneapolis: University of Minnesota Press, 2019.

Howell, Elizabeth. "James Irwin: Eighth Man on the Moon." Space.com, April 8, 2013. https://www.space.com/20567-james-irwin-apollo-15-astronaut.html.

Hsu, Jeremy. "The Myth of America's Love Affair with the Moon." Space.com, January 13, 2011. https://www.space.com/10601-apollo-moon-program-public-support-myth.html.

"Is the Earth Safe from Lunar Contamination?" *Time*, June 13, 1969.

James, Sarah. "We Are the Ones Who Have Everything to Lose." In *Arctic Voices: Resistance at the Tipping Point*, edited by Subhankar Banerjee, 260–66. New York: Seven Stories, 2013.

Jeffers, Robinson. "The Treasure." In *The Selected Poetry of Robinson Jeffers*, edited by Tim Hunt, 100. New York: Random House, 1959.

Johnson, Clarence Sholé. "Reading Between the Lines: Kathryn Gines on Hannah Arendt and Antiblack Racism." *Southern Journal of Philosophy* 47 (2009): 77–83.

Johnson, Jenell. "Disability, Animals, and the Rhetorical Boundaries of Personhood." *JAC* 32, nos. 1–2 (2012): 372–82.

Johnson, Sarah Stewart, Martin B. Hebsgaard, Torben R. Christensen, Mikhail Mastepanov, Rasmus Nielsen, Kasper Munch, Tina Brand, et al. "Ancient Bacteria Show Evidence of DNA Repair." *Proceedings of the National Academy of Sciences* 104, no. 36 (2007): 14401–5.

Jonas, Hans. *The Phenomenon of Life: Toward a Philosophical Biology*. Evanston: Northwestern University Press, 2001.

Jones, Donna V. *The Racial Discourses of Life Philosophy: Négritude, Vitalism, and Modernity*. New York: Columbia University Press, 2010.

Jordan, John W. "Kennedy's Romantic Moon and Its Rhetorical Legacy for Space Exploration." *Rhetoric and Public Affairs* 6, no. 2 (2003): 209–31.

Kalanithi, Paul. *When Breath Becomes Air*. New York: Random House, 2016.

Kanas, Nick and Dietrich Manzey. *Space Psychology and Psychiatry*. Dordrecht: Springer, 2008.

Kant, Immanuel. *Critique of Judgment*. Translated by James Creed Meredith. Oxford: Oxford University Press, 2007.

———. *Universal Natural History and Theory of Heaven*. Oxford: Oxford Text Archive. http://hdl.handle.net/20.500.12024/2521.

Keim, Brandon. "Be a Parent of a Brand-New Word: Astrobioethics." *Wired*, August 5, 2008. http://www.wired.com/2008/08/wired-science-c.

Keller, David R. "Deep Ecology." In *Encyclopedia of Environmental Ethics and Philosophy*, edited by J. Baird Callicott and Robert Frodeman, 206–11. Detroit: Macmillan Reference, 2008.

Keränen, Lisa. "Addressing the Epidemic of Epidemics: Germs, Security, and a Call for Biocriticism." *Quarterly Journal of Speech* 97, no. 2 (2011): 224–44.

Killingsworth, Jimmie, and Jacqueline S. Palmer. "Millennial Ecology: The Apocalyptic Narrative from Silent Spring to Global Warming." In *Green Culture: Environmental Rhetoric in Contemporary America*, edited by Carl G. Herndl and Stuart C. Brown, 21–45. Madison: University of Wisconsin Press, 1996.

Kiminek, Gerhard, Catharine Conley, Victoria Hipkin, and Hajime Yano. "COSPAR's Planetary Protection Policy." *Space Research Today* 200 (December 2017): 12–25.

Kimmerer, Robin Wall. *Braiding Sweetgrass: Indigenous Wisdom, Scientific Knowledge, and the Teachings of Plants*. Minneapolis: Milkweed Editions, 2013.

Kollin, Susan. *Nature's State: Imagining Alaska as the Last Frontier*. Chapel Hill: University of North Carolina Press, 2001.

Krabbe, Erik C. W. "Arne Næss (1912–2009)." *Argumentation* 24, no. 4 (2010): 527–30.

Langer, Julian. "Interview with Gail Bradbrook." *Engaged Dharma*, July 30, 2018. https://engagedharma.net/2018/11/24/extinction-rebellion.

Lazier, Benjamin. "Earthrise; or, The Globalization of the World Picture." *American Historical Review* 116, no. 3 (2011): 602–30.

Lebovic, Nitzan. *The Philosophy of Life and Death: Ludwig Klages and the Rise of a Nazi Biopolitics*. New York: Palgrave, 2013.

Lederberg, Joshua. "Exobiology." Letter. *Science* 142 (1963): 1126.

———. "Exobiology: Approaches to Life Beyond the Earth." *Science* 132, no. 3424 (August 1960): 393–400.

———. "Signs of Life." *Nature* 207, no. 4992 (1965): 9–13.

———. "Sputnik + 30." *Journal of Genetics* 66, no. 3 (1987): 217–20.

Lederberg, Joshua, and Dean B. Cowie. "Moondust." *Science* 127, no. 3313 (1958): 1473–75.

Legagneux, Pierre, Nicolas Casajus, Kevin Cazelles, Clément Chevallier, Marion Chevrinais, Lorelei Guéry, Claire Jacquet, et al. "Our House Is Burning: Discrepancy in Climate Change vs. Biodiversity Coverage in the Media as Compared to Scientific Literature." *Frontiers in Ecology and Evolution* 5, no. 175 (2018): 1–6.

Lehmann, Ernst. *Biologischer Wille: Wege und Ziele Biologischer Arbeit im Neuen Reich*. München: J.F. Lehmann, 1934.

Leopold, Aldo. *A Sand County Almanac and Sketches Here and There*. London: Oxford University Press, 1949.

Lewiecki-Wilson, Cynthia. "Ableist Rhetorics, Nevertheless: Disability and Animal Rights in the Work of Peter Singer and Martha Nussbaum." *JAC* 31, nos. 1–2 (2011): 71–101.

Lewis, Martin. "Toward a Self-Critical Environmentalism." *Politics and the Life Sciences* 18, no. 2 (1999): 229–31.

Lidgard, Scott, and Lynn K. Nyhart. Introduction to *Biological Individuality: Integrating Scientific, Philosophical, and Historical Perspectives*, edited by Scott Lidgard and Lynn K. Nyhart, 1–16. Chicago: University of Chicago Press, 2017.

Lind, Katherine Dominique. "The Unbearable Loss of Beings: Curating, Documenting, and Resisting Anthropogenic Mass Extinction." PhD diss., Indiana University, 2018.

Linkola, Pentti. *Can Life Prevail? A Revolutionary Approach to the Climate Crisis*. Budapest: Arktos Media, 2009.

Lukes, Steven. *Individualism*. Essex: ECPR Press, 2006.

Lupisella, Mark. "The Rights of Martians." *Space Policy* 13, no. 2 (1997): 89–94.

Lyne, John. "Bio-rhetorics: Moralizing the Life Sciences." In *The Rhetorical Turn: Invention and Persuasion in the Conduct of Inquiry*, edited by Herbert W. Simons, 35–57. Chicago: University of Chicago Press, 1990.

Machery, Edouard. "Why I Stopped Worrying About the Definition of Life . . . and Why You Should as Well." *Synthese* 185, no. 1 (2012): 145–64.

Mack, Ashley Noel, and Tiara R. Na'puti. "'Our Bodies Are Not Terra Nullius': Building a Decolonial Feminist Resistance to Gendered Violence." *Women's Studies in Communication* 42, no. 3 (2019): 347–70.

Mangold, Eli, and Charles Goehring. "Identification by Transitive Property: Intermediated Consubstantiality in the NFL's Salute to Service Campaign." *Critical Studies in Media Communication* 35, no. 5 (2018): 503–16.

Mann, Thomas. *The Magic Mountain*. Translated by John E. Woods. New York: Everyman's Library, 2005. First published 1924.

Margulis, Lynn, and Dorion Sagan. *What Is Life?* Berkeley: University of California Press, 2000.

Mariscal, Carlos, and W. Ford Doolittle. "Life and Life Only: A Radical Alternative to Life Definitionism." *Synthese* 197 (2018): 2975–89.

Markley, Robert. *Dying Planet: Mars in Science and the Imagination*. Durham: Duke University Press, 2005.

Marra, Peter P., and Chris Santella. *Cat Wars: The Devasting Consequences of a Cuddly Killer*. Princeton: Princeton University Press, 2016.

Marshall, Alan. "Ethics and the Extraterrestrial Environment." *Journal of Applied Philosophy* 10, no. 2 (1993): 227–36.

Mbembe, Achille. *Critique of Black Reason*. Translated by Laurent Dubois. Durham: Duke University Press, 2017.

———. *Necropolitics*. Translated by Steven Corcoran. Durham: Duke University Press, 2019.

———. *Out of the Dark Night: Essays on Decolonization*. New York: Columbia University Press, 2021.

———. "The Universal Right to Breathe." Translated by Carolyn Sheard. *Inquiry* 47, no. S2 (2021): S58–S62.

McKay, Christopher P. "Biologically Reversible Exploration." *Science* 323 (2009): 718.

———. "Does Mars Have Rights? An Approach to the Environmental Ethics of Planetary Engineering." *Moral Expertise*, edited by Don MacNiven, 184–97. London: Routledge, 1990.

———. "Planetary Ecosynthesis on Mars: Restoration Ecology and Environmental Ethics." In *Exploring the Origin, Extent, and Future of Life: Philosophical, Ethical, and Theological Perspectives*, edited by C. Bertka, 245–60. Cambridge: Cambridge University Press, 2009.

McKay, David S., Everett K. Gibson, Kathie L. Thomas-Keprta, Hojatollah Vali, Christopher S. Romanek, Simon J. Clemett, Xavier D. F. Chillier, Claude R. Maechling, and Richard N. Zare. "Search for Past Life on Mars: Possible Relic Biogenic Activity in Martian Meteorite ALH84001." *Science* 273, no. 5277 (1996): 924–30.

Meadows, Donella. "Chicken Little, Cassandra, and the Real Wolf: So Many Ways to Think About the Future." *Whole Earth*, Spring 1999, 106–11.

Meltzer, Michael. *When Biospheres Collide: A History of NASA's Planetary Protection Programs*. Washington, DC: Government Printing Office, 2012.

Messeri, Lisa. *Placing Outer Space: An Earthly Ethnography of Other Worlds*. Durham: Duke University Press, 2016.

Metildi, Nicole. "Water Is Life: Illuminating Slow Violence Through Coalitional Movements and Narratives." Paper presented at Waterlines, the Conference on Communication and Environment, Vancouver, Canada, June 17–21, 2019. https://theieca.org/sites/default/files/COCE2019/program/s223.html.

Mignolo, Walter. *The Darker Side of Western Modernity: Global Futures, Decolonial Options*. Durham: Duke University Press, 2011.

Miller, Carolyn R., and Molly Hartzog. "'Tree Thinking': The Rhetoric of Tree Diagrams in Biological Thought." *Poroi: An Interdisciplinary Journal of Rhetorical Analysis and Invention* 15, no. 2 (2020): art. 2.

Mitchell, Robert. *Experimental Life: Vitalism in Romantic Science and Literature*. Baltimore: Johns Hopkins Press, 2013.

Mody, Cyrus. "A Little Dirt Never Hurt Anyone: Knowledge-Making and Contamination in Materials Science." *Social Studies of Science* 31, no. 1 (2001): 7–36.

Mortimer-Sandilands, Catriona, and Bruce Erickson. *Queer Ecologies: Sex, Nature, Politics, Desire*. Bloomington: Indiana University Press, 2010.

Murphy, Michelle. *The Economization of Life*. Durham: Duke University Press, 2017.

Murray, Jeffrey W. "An Other Ethics for Kenneth Burke." *Communication Studies* 49, no. 1 (1998): 29–48.

Murray, Stuart J. "Affirming the Human? The Question of Biopolitics." *Law, Culture and the Humanities* 12, no. 3 (2016): 485–95.

———. "Aporia: Towards an Ethic of Critique." *Aporia* 1, no. 1 (2009): 8–14.

Næss, Arne. *Ecology, Community and Lifestyle: Outline of an Ecosophy*. Cambridge: Cambridge University Press, 1990.

———. *The Ecology of Wisdom: Writings by Arne Næss*, edited by Alan Drengson and Bill Devall. Berkeley: Counterpoint, 2008.

———. *Life's Philosophy: Reason and Feeling in a Deeper World*. Translated by Roland Huntford. Athens: University of Georgia Press, 2002.

———. "'Man Apart' and Deep Ecology: A Reply to Reed." *Environmental Ethics* 12, no. 2 (1990): 185–92.

———. *The Selected Works of Arne Næss*. Vols. 1–10. Edited by Harold Glasser and Alan Drengson. Dordrecht: Springer, 2005.

———. "Self-Realization: An Ecological Approach to Being in the World." In *Deep Ecology for the Twenty-First Century*, edited by George Sessions, 225–39. Boston and London: Shambala, 1995.

———. "The Shallow and the Deep, Long-Range Ecology Movement: A Summary." *Inquiry* 16 (1973): 95–100.

Na'puti, Tiara R. "Speaking of Indigeneity: Navigating Genealogies Against Erasure and #RhetoricSoWhite." *Quarterly Journal of Speech*, 105, no. 4 (2019): 495–501.

Nash, Roderick Frazier. *The Rights of Nature: A History of Environmental Ethics*. Madison: University of Wisconsin Press, 1989.

National Library of Medicine. "Launching a New Science: Exobiology and the Exploration of Space." Joshua Lederberg Papers. Accessed January 13, 2021. https://profiles.nlm.nih.gov/spotlight/bb/feature/launching-a-new-science-exobiology-and-the-exploration-of-space.

National Public Radio. "Science Fiction, Sagan Message, Headed to Mars." *All Things Considered*, August 4, 2007. https://www.npr.org/templates/story/story.php?storyId=12507759.

Nealon, Jeffrey. *Plant Theory: Biopower and Vegetable Life*. Stanford: Stanford University Press, 2020.

Nyhart, Lynn K., and Scott Lidgard. "Individuals at the Center of Biology: Rudolf Leuckart's *Polymorphismus der Individuen* and the Ongoing Narrative of Parts and Wholes." *Journal of the History of Biology* 44, no. 3 (2011): 373–443.

Oelschlaeger, Max. *The Idea of Wilderness: From Prehistory to the Age of Ecology*. New Haven: Yale University Press, 1991.

Oliver, Kendrick. *To Touch the Face of God: The Sacred, the Profane, and the American Space Program, 1957–1975*. Baltimore: Johns Hopkins University Press, 2013.

Olson, Christa J. *American Magnitude: Hemispheric Vision and Public Feeling in the United States*. Columbus: Ohio State University Press, 2021.

———. *Constitutive Visions: Indigeneity and Commonplaces of National Identity in Republican Ecuador*. University Park: Pennsylvania State University Press, 2013.

Oravec, Christine. "John Muir, Yosemite, and the Sublime Response: A Study in the Rhetoric of Preservationism." *Quarterly Journal of Speech* 67, no. 3 (1981): 245–58.

Ore, Ersula. *Lynching: Violence, Rhetoric, and American Identity*. Oxford: University of Mississippi Press, 2019.

Osborne, Thomas. "Vitalism as Pathos." *Biosemiotics* 9, no. 2 (2016): 185–205.

Owen, Tess. "How a Small Budapest Publishing House Is Quietly Fueling Far-Right Extremism." *Vice*, May 30, 2019. https://www.vice.com/en/article/3k3558/how-a-small-budapest-publishing-house-is-quietly-fueling-far-right-extremism.

Peters, Tori Thompson. "My Body, My Cells: Rhetoric and the Molecularization of the Human." *Rhetoric Society Quarterly* 51, no. 2 (2021): 123–37.

Pezzullo, Phaedra. "Unearthing the Marvelous: Environmental Imprints on Rhetorical Criticism." *Review of Communication* 16, no. 1 (2016): 25–42.

Pezzullo, Phaedra, and Robert Cox. *Environmental Communication and the Public Sphere*. 5th ed. New York: Sage Books, 2017.

Pezzullo, Phaedra, and Ronald Sandler. "Introduction: Revisiting the Environmental Justice to Environmentalism." In *Environmental Justice and Environmentalism*, edited by Ronald Sandler and Phaedra C. Pezzullo, 1–24. Cambridge: MIT Press, 2007.

Pimm, Stuart L., Clinton N. Jenkins, Robin Abell, Thomas M. Brooks, John L. Gittleman, Lucas N. Joppa, Peter H. Raven, Callum M. Roberts, and Joseph O. Sexton. "The Biodiversity of Species and Their Rates of Extinction, Distribution, and Protection." *Science* 344, no. 6187 (2014): 1246752.

Planetary Collective. *Overview*. Short film. Directed by Guy Reid. December 2012. http://weareplanetary.com/overview-short-film.

Plumwood, Val. "Deep Ecology, Deep Pockets, and Deep Problems." In *Beneath the Surface: Critical Essays in the Philosophy of Deep Ecology*, edited by Eric Katz, Andrew Light, and David Rothenberg, 59–84. Cambridge: MIT Press, 2000.

Prasch, Allison M. "Toward a Rhetorical Theory of Deixis." *Quarterly Journal of Speech* 102, no. 2 (2016): 166–93.

Pravinchandra, Shital. "One Species, Same Difference? Postcolonial Critique and the Concept of Life." *New Literary History* 47, no. 1 (2016): 27–48.

Protopapadakis, Evangelos D. "Environmental Ethics and Linkola's Ecofascism: An Ethics Beyond Humanism." *Frontiers of Philosophy in China* 9, no. 4 (2014): 586–601.

Puar, Jasbir K. *Terrorist Assemblages: Homonationalism in Queer Times*. Durham: Duke University Press, 2018.

Querejazu, Amaya. "Encountering the Pluriverse: Looking for Alternatives in Other Worlds." *Revista Brasileira de Política Internacional* 59, no. 2 (2016).

Ramírez, J. Jesse. "From Woking to New York; or, American Wars: A Very Brief Cultural History of *War of the Worlds* in North American Translation." Paper presented at the H. G. Wells Society Annual Conference, Woking, UK, July 8–10, 2016.

Randolph, Richard O., and Christopher P. McKay. "Protecting and Expanding the Richness and Diversity of Life, an Ethic of Astrobiology Research and Space Exploration." *International Journal of Astrobiology* 13, no. 1 (2014): 28–34.

Randolph, Richard O., Margaret S. Race, and Christopher P. McKay. "Reconsidering the Theological and Ethical Implications of Extraterrestrial Life." *CTNS Bulletin* 17, no. 3 (1997): 1–8.

Raymont, Henry. "Publishers Hitching Star to the Moon Expedition." *New York Times*, July 16, 1969, 30.

Redd, Nola Taylor. "Planetary Protection: Contamination Debate Still Simmers." Space.com, May 8, 2017. https://www.space.com/36708-planetary-protection-astrobiology-nasa-missions.html.

Reed, Peter, "Man Apart: An Alternative to the Self-Realization Approach." *Environmental Ethics* 11, no. 1 (1989): 53–69.

Rittel, H. W., and M. M. Webber. "Dilemmas in a General Theory of Planning." *Policy Sciences* 4, no. 2 (1973): 155–69.

Rivers, Nathaniel A. "Deep Ambivalences and Wild Objects: Toward a Strange Rhetoric." *Rhetoric Society Quarterly* 45, no. 5 (2015): 420–40.

Robinson, Kim Stanley. *Green Mars*. New York: Del Rey, 2017. First published 1993.

———. *Red Mars*. New York: Del Rey, 2017. First published 1992.

Rose, Nikolas. *The Politics of Life Itself*. Princeton: Princeton University Press, 2009.

Rosen, Robert. *Life Itself: A Comprehensive Inquiry into the Nature, Origin, and Fabrication of Life*. New York: Columbia University Press, 1991.

Rowland, Allison. *Zoetropes and the Politics of Humanhood*. Columbus: Ohio University Press, 2020.

Rowlands, Mark. *Can Animals Be Moral?* Oxford: Oxford University Press, 2012.

Rummel, John D. "Planetary Exploration in the Time of Astrobiology: Protecting Against Biological Contamination." *Proceedings of the National Academy of Sciences* 98, no. 5 (2001): 2128–31.

Rummel, John D., and Linda Billings. "Issues in Planetary Protection: Policy, Protocol and Implementation." *Space Policy* 20, no. 1 (2004): 49–54.

Sackey, Donnie Johnson, Casey Boyle, Mai Nou Xiong, Gabriela Raquel Ríos, Kristin L. Arola, and Scot Barnett. "Perspectives on Cultural and Posthumanist Rhetorics." *Rhetoric Review* 38, no. 4 (2019): 375–401.

Sagan, Carl. *Cosmos*. New York: Ballantine Books, 2013.

Sagan, Dorion. "What Narcissus Saw: The Oceanic 'Eye.'" In *Slanted Truths: Essays on Gaia, Symbiosis, and Evolution*, edited by Lynn Margulis and Dorion Sagan, 185–200. New York: Copernicus, 1997.

Salas, Dominique. "Decolonizing Exigency: Settler Exigences in the Wisconsin Winnebago Mission Home." *Rhetoric Society Quarterly* 52, no. 2: 107–21.

Samuels, Ellen J. *Fantasies of Identification: Disability, Gender, Race*. New York: New York University Press, 2014.

Schiebinger, Londa. *Nature's Body: Gender in the Making of Modern Science*. Boston: Beacon, 1993.

Schmeck, Harold M. "Rockets' Effect on Moon Feared: Non-Sterile Space Ship May Distort Life on Other Planets, Experts Say." *New York Times*, July 6, 1958.

Schott, Webster. Review of *The Andromeda Strain*, by Michael Crichton. *New York Times*, June 8, 1969, BR4.

Schweitzer, Albert. *Reverence for Life: The Ethics of Albert Schweitzer for the Twenty-First Century*. Edited by Marvin Meyer and Kurt Bergel. Syracuse: Syracuse University Press, 2002.

Seed, David. "The Course of Empire: A Survey of the Imperial Theme in Early Anglophone Science Fiction." *Science Fiction Studies* 37, no. 2 (2010): 230–52.

Sessions, George, and Arne Næss. "The Basic Principles of Deep Ecology." *Earth First!* 4, no. 6 (1984). Republished by the Environment & Society Portal, Multimedia Library. http://www.environmentandsociety.org/node/6853.

Sharpe, Christina. *In the Wake: On Blackness and Being*. Durham: Duke University Press, 2016.

Sheehan, William, and Thomas Dobbins. "The Spokes of Venus: An Illusion Explained." *Journal for the History of Astronomy* 34, no. 1 (2003): 53–63.

Shome, Raka. "Postcolonial Interventions in the Rhetorical Canon: An 'Other' View," *Communication Theory* 6, no. 1 (1996): 40–59.

Simaika, John P., and Michael J. Samways. "Biophilia as a Universal Ethic for Conserving Biodiversity." *Conservation Biology* 24, no. 3 (2010): 903–6.

Smith, Kelly C. "The Curious Case of the Martian Microbes: Mariomania, Intrinsic Value and the Prime Directive." In *The Ethics of Space Exploration*, edited by James Swartz and Tony Milligan, 195–208. New York: Springer, 2016.

Sowards, Stacy. "Identification Through Orangutans: Destabilizing the Nature/Culture Dualism." *Ethic and the Environment* 11, no. 2 (2006): 45–61.

Squier, Susan Merrill. *Liminal Lives: Imagining the Human at the Frontiers of Biomedicine*. Durham: Duke University Press, 2004.

Steele, Edward J., Shirwan Al-Mufti, Kenneth A. Augustyn, Rohana Chandrajith, John P. Coghlan, S. G. Coulson, Sudipto Ghosh, et al. "Cause of Cambrian Explosion—Terrestrial or Cosmic?" *Progress in Biophysics and Molecular Biology* 136 (2018): 3–23.

Stone, Robert, dir. *Chasing the Moon*. Part 3. Documentary. PBS, July 2019. https://www.pbs.org/wgbh/americanexperience/films/chasing-moon.

Stormer, Nathan. *Articulating Life's Memory: US Medical Rhetoric About Abortion in the Nineteenth Century*. New York: Lexington Books, 2002.

Strick, James E. "Creating a Cosmic Discipline: The Crystallization and Consolidation of Exobiology, 1957–1973." *Journal of the History of Biology* 37, no. 1 (2004): 131–80.

Subramaniam, Banu. "The Aliens Have Landed! Reflections on the Rhetoric of Biological Invasions." *Meridians: Feminism, Race, Transnationalism* 2, no. 1 (2001): 26–40.

Sullivan, Walter. "Guarding Against Moon Bugs." *New York Times*, June 15, 1969, E8.

TallBear, Kim. "Beyond the Life/Not-Life Binary: A Feminist-Indigenous Reading of Cryopreservation, Interspecies Thinking and the New Materialisms." In *Cryopolitics: Frozen Life in a Melting World*, edited by Joanna Radin and Emma Cowal, 179–202. Cambridge: MIT Press, 2017.

———. "An Indigenous Reflection on Working Beyond the Human/Not Human." *GLQ: A Journal of Lesbian and Gay Studies* 21, nos. 2–3 (2015): 230–35.

Tarantola, Andrew. "New Mars Meteorite Is Selling for $22,500 Per Ounce—Ten Times the Price of Gold." *Gizmodo*, January 18, 2012. https://gizmodo.com/new-mars-meteorite-is-selling-for-22-500-per-ounce-10-5877062.

Taylor, Charles. *Sources of the Self: The Making of the Modern Identity*. Cambridge: Harvard University Press, 1992.

Taylor, Keeanga-Yamahtta. *From #BlackLivesMatter to Black Liberation*. Chicago: Haymarket Books, 2016.

Taylor, Paul W. *Respect for Nature: A Theory of Environmental Ethics*. Princeton: Princeton University Press, 2011. First published 1986.

Temple, Robert. "The Prehistory of Panspermia: Astrophysical or Metaphysical?" *International Journal of Astrobiology* 6, no. 2 (2007): 169–80.

Theophrastus. *Enquiry into Plants*. Book 3. Loeb Classical Library 1. Translated by A. Hort. Cambridge: Harvard University Press, 1990.

These Exit Times. Nos. 1–4. Portland, OR: Les U. Knight, 1991–94.

Todd, Zoe. "Fish Pluralities: Human-Animal Relations and Sites of Engagement in Paulatuuq, Arctic Canada." *Études/Inuit/Studies* 38, nos. 1–2 (2014): 217–38.

Toulmin, Stephen. *The Return to Cosmology: Postmodern Science and the Theology of Nature*. Berkeley: University of California Press, 1982.

Twilley, Nicola. "Meet the Martians." *New Yorker*, October 8, 2015. https://www.newyorker.com/tech/annals-of-technology/meet-the-martians.

Uekotter, Frank. *The Green and the Brown: A History of Conservation in Nazi Germany*. Cambridge: Cambridge University Press, 2006.

van Dooren, Thom. *Flight Ways: Life and Loss and the End of Extinction*. New York: Columbia University Press, 2014.

Vatter, Miguel. "Natality and Biopolitics in Hannah Arendt." *Revista de Ciencia Política* 26, no. 2 (2006): 137–59.

Veldman, Robin Globus. "Narrating the Environmental Apocalypse: How Imagining the End Facilitates Moral Reasoning Among Environmental Activists." *Ethics and the Environment* 17, no. 1 (2012): 1–23.

Vernadsky, Vladimir I. *The Biosphere*. Translated by David B. Langmuir. New York: Copernicus, 1998.

Vitalist International. "Life Finds a Way." *Commune*, Winter 2020. https://communemag.com/life-finds-a-way.

Vizenor, Gerald Robert. *Manifest Manners: Narratives on Postindian Survivance*. Lincoln: University of Nebraska Press, 1999.

Wald, Priscilla. *Contagious: Cultures, Carriers, and the Outbreak Narrative*. Durham: Duke University Press, 2008.

Walsh, Denis M. "Objectcy and Agency: Towards a Methodological Vitalism." In *Everything Flows: Towards a Processual Philosophy of Biology*, edited by Daniel J. Nicholson and John Dupré, 167–85. Oxford: Oxford University Press, 2018.

———. *Organisms, Agency, and Evolution*. Cambridge: Cambridge University Press, 2015.

Wanzer, Darrel. "Delinking Rhetoric, or Revisiting McGee's Fragmentation Thesis Through Decoloniality," *Rhetoric and Public Affairs* 15, no. 4 (2012): 647–57.

Watson, Annette, and Orville H. Huntington. "They're Here—I Can Feel Them: The Epistemic Spaces of Indigenous and Western Knowledges." *Social and Cultural Geography* 9, no. 3 (2008): 257–81.

Watts, Jonathan. "Stop Biodiversity Loss or We Could Face Our Own Extinction, Warns UN." *Washington Post*, October 15, 2018.

———. "We Have 12 Years to Limit Climate Change Catastrophe, Warns UN." *The Guardian*, October 8, 2018. https://www.theguardian.com/environment/2018/oct/08/global-warming-must-not-exceed-15c-warns-landmark-un-report.

Watts, Vanessa. "Indigenous Place-Thought and Agency Amongst Humans and Non-humans (First Woman and Sky Woman Go on a European World Tour!)." *Decolonization: Indigeneity, Education and Society* 2, no. 1 (2013): 20–34.

Weisman, Alan. *The World Without Us*. New York: Picador, 2008.

Wells, H. G. *The War of the Worlds*. London: William Heinemann, 1898.

———. "War of the Worlds: Installment 9." *Pearson's*, December 1897, 736–45.

Welter, Volker M. "From Disc to Sphere: Taking on the Whole Earth." *Cabinet* 40 (2011): 19–25.

White, Frank. *The Overview Effect: Space Exploration and Human Evolution*. Boston: Houghton Mifflin, 1987.

Whyte, Kyle. "Against Crisis Epistemology." In *Handbook of Critical Indigenous Studies*, edited by Brendan Hokowhitu, Aileen Moreton-Robinson, Linda Tuhiwai-Smith, Steve Larkin, and Chris Andersen, 52–64. London: Routledge, 2021.

———. "The Dakota Access Pipeline, Environmental Injustice, and U.S. Colonialism." *RED INK: An International Journal of Indigenous Literature, Arts, and Humanities*: 19, no. 1 (2017): 154–69.

———. "Our Ancestors' Dystopia Now: Indigenous Conservation and the Anthropocene." In *Routledge Companion to the Environmental Humanities*, edited by Ursula Heise, Jon Christense, and Michelle Niemann, 208–15. London: Routledge, 2016.

———. "Way Beyond the Lifeboat: An Indigenous Allegory of Climate Justice." In *Climate Futures: Re-imagining Global Climate Justice*, edited by Kum-Kum Bhavani, John Foran, Priya A. Kurian, and Debashish Munshi, 11–20. London: Bloomsbury Academic, 2019.

Wilford, John Noble. "Clues in Meteorite Seem to Show Signs of Life on Mars Long Ago." *New York Times*, August 7, 1996. https://www.nytimes.com/1996/08/07/us/clues-in-meteorite-seem-to-show-signs-of-life-on-mars-long-ago.html.

Wilson, Edmund O. *Biophilia*. Cambridge: Harvard University Press, 1984.

Wilson, Stan. "Self-as-Relationship in Indigenous Research." *Canadian Journal of Native Education* 25, no. 2 (2001): 91–92.

Woese, Carl R., Otto Kandler, and Mark L. Wheelis. "Towards a Natural System of Organisms: Proposal for the Domains Archaea, Bacteria, and Eucarya." *Proceedings of the National Academy of Sciences* 87, no. 12 (1990): 4576–79.

Wolfe, Audra J. "Germs in Space: Joshua Lederberg, Exobiology, and the Public Imagination, 1958–1964." *Isis* 93 (2002): 183–205.

Wretched of the Earth. "An Open Letter to Extinction Rebellion." *Common Dreams*, May 4, 2019. https://www.commondreams.org/views/2019/05/04/open-letter-extinction-rebellion.

Wynter, Sylvia. "Unsettling the Coloniality of Being/Power/Truth/Freedom: Towards the Human After Man, Its Overrepresentation – An Argument." *CR: The New Centennial Review* 3, no. 3 (2003): 257–337.

Wynter, Sylvia, and Katherine McKittrick. "Unparalleled Catastrophe for Our Species? Or to Give Humanness a Different Future: Conversations." In *Sylvia Wynter: On Being Human as Praxis*, edited by Katherine McKittrick, 9–89. Durham: Duke University Press, 2015.

Yaden, David B., Jonathan Iwry, Kelley J. Slack, Johannes C. Eichstaedt, Yukun Zhao, George E. Vaillant, and Andrew B. Newberg. "The Overview Effect: Awe and Self-Transcendent Experience in Space Flight." *Psychology of Consciousness: Theory, Research, and Practice* 3, no. 1 (2016): 1–11.

Zubrin, Robert. *The Case for Mars: The Plan to Settle the Red Planet and Why We Must.* New York: Free Press, 2011.

———. "The Case for Terraforming Mars." In *On to Mars: Colonizing a New World*, edited by Robert Zubrin and Frank Crossman, 179–80. Ontario: Apogee Books, 2005.

Index